# Animal Models in Medical Research

*Edited by Pınar Atukeren*

Published in London, United Kingdom

Animal Models in Medical Research
http://dx.doi.org/10.5772/intechopen.1003495
Edited by Pınar Atukeren

Contributors
Ana Luisa Gutierréz-Valdez, Anfeng Xiang, Anna Soloveva, Beatriz Gomes Pinto, Camilla Arvinius, Cesar Alfonso Garcia Caballero, Clarissa Perdigão Melo Ferraz, Claudia Dorado-Martínez, Cruz Rodriguez-Bobada, David Anuoluwapo Oyeniran, Eleftheria Mavrigiannaki, Enrique Montiel-Flores, Fabiano Marcelo Fabris, Fernando Marco, Geraldo de França Júnior, Gideon Olamilekan Oluwatunase, Ioannis Georgopoulos, Javier Sanchez-Betancourt, Jesús Espinosa-Villanueva, José Luis Ordoñez-Librado, Leonardo Reynoso-Erazo, Lumei Xiang, Maria Rosa Avila-Costa, Olayinka Olawale Asafa, Onome Bright Ogenetega, Orleâncio Gomes Ripardo de Azevedo, Patricia Aley-Medina, Patrick Godwin Okwute, Petr Peretyagin, Radamés Bezerra Melo, Rocío Tron-Alvarez, Tadeja Kuret, Vianey Rodríguez-Lara, Virginia Ponz-Lueza, Xiaoqing Tan, Xiaoxia Wang, Yaiza Lopiz, Zhichao Wang



First published in London, United Kingdom, 2025 by IntechOpen
IntechOpen is the global imprint of INTECHOPEN LIMITED, registered in England and Wales, registration number: 11086078, 167-169 Great Portland Street, London, W1W 5PF, United Kingdom

For EU product safety concerns: IN TECH d.o.o., Prolaz Marije Krucifikse Kozulić 3, 51000 Rijeka, Croatia, info@intechopen.com or visit our website at intechopen.com.

British Library Cataloguing-in-Publication Data
A catalogue record for this book is available from the British Library

Animal Models in Medical Research
Edited by Pınar Atukeren
p. cm.
Print ISBN 978-0-85014-930-2
Online ISBN 978-0-85014-929-6
eBook (PDF) ISBN 978-0-85014-931-9

# Meet the editor

Dr. Pınar Atukeren is the Dean of the Faculty of Pharmacy at Istanbul University-Cerrahpaşa, Istanbul, Turkey. She is also teaching in the Department of Medical Biochemistry of Cerrahpaşa Medical Faculty, in the same university, as a senior professor from where she obtained her Ph.D. in 2005. She had been awarded the National Research Grant for Successful Students from the Scientific and Technical Research Council of Turkey. Her research interests are oxidative and nitrosative stress, free radicals and antioxidants, natural therapeutics, neurodegenerative diseases, brain tumors, and aging. Dr. Atukeren has published many papers in indexed journals and edited and authored both international and national books and book chapters. She is a reviewer for many international peer-reviewed journals. She has attended many conferences and workshops and is involved in many training courses and consultations. She is also the head of the Animal Welfare and Ethics Committee and a member of the Scientific Research Projects Coordination Unit of Istanbul University-Cerrahpaşa. At the community level, Dr. Atukeren is an active member of several governmental and non-governmental societies and committees.

# Contents

# Preface

Medical research has long relied on experimental animal models to explore disease mechanisms, develop new treatments, and ensure the safety of medical interventions before human trials. These models are indispensable tools for understanding complex biological processes in a controlled environment, providing valuable insights that cannot be obtained through in vitro studies alone.

Using animals in research has led to groundbreaking discoveries in fields such as pharmacology, oncology, neurology, and immunology. From the development of vaccines and antibiotics to advances in organ transplantation and cancer therapies, experimental animal models have played a crucial role in shaping modern medicine. However, their use also raises ethical concerns, necessitating strict regulations, ethical review boards, and the application of the "3Rs" principle—Replacement, Reduction, and Refinement—to minimize animal suffering while maximizing scientific benefit.

A well-designed experimental animal model needs an understanding of differences and similarities between the responses of both humans and animals, incorporating this into the aims of the study. Most of the therapeutic molecules and drugs are developed or validated via animal experiments in current medicine, yet ensuring that the experimental research is in accordance with animal welfare and ethical rules should be the main point.

This book aims to provide an overview of the significance, applications, and ethical considerations surrounding various experimental animal models in medical research. By examining different model organisms, their strengths and limitations, and the evolving landscape of alternative methodologies, we seek to highlight the delicate balance between scientific progress and ethical responsibility.

Hopefully, the experimental animal models discussed in this book will contribute to a deeper understanding of the role of animal models in medical research and inspire further advancements in both biomedical sciences and ethical research practices.

**Dr. Pınar Atukeren**
Professor,
Faculty of Pharmacy,
Istanbul University-Cerrahpaşa,
Istanbul, Türkiye

Cerrahpasa Medical Faculty,
Department of Biochemistry,
Istanbul, Türkiye

Chapter 1

# Acute and Chronic Rotator Cuff Tear: Experimental Rat Model

*Virginia Ponz-Lueza, Yaiza Lopiz, Camilla Arvinius, Cruz Rodriguez-Bobada and Fernando Marco*

## Abstract

This chapter describes the injury and repair of the supraspinatus tendon in a rat. This model can be applied to produce both acute and a chronic injury. In acute injury, the tendon section and repair are performed at the same time, under general anaesthesia. In chronic injury, in the first surgical intervention, the supraspinatus tendon repair is performed with local anaesthesia and sedation. Four weeks after the injury, the injury is considered chronic, and we proceed to perform tendon repair under general anaesthesia. Depending on what we want to evaluate in both models after tendon repair, the model is variable and can be adapted to the contribution of stem cells in suspension or to the use of scaffolds together with stem cells or growth factors. Four months after the tendon repair, the animal is euthanized in order to perform biomechanical and histological studies of the tendon structure.

**Keywords:** rotator cuff tear, supraspinatus, rat model, scaffold, mesenchymal stem cells, acute injury, chronic injury

## 1. Introduction

Chronic rotator cuff tears affect approximately 30% of the population over 60 years of age, many of them symptomatic limiting the basic activities of daily living in our patients [1].

Arthroscopic surgical treatment of these lesions is the most commonly used treatment [2]. However, despite improvements in repair techniques for rotator cuff injuries, the rate of retears continues to be very high, with percentages ranging between 11 and 94%, depending on the according to other authors [3, 4].

In order to improve the biological healing process, scientific studies are focused on the implementation of different techniques and alternatives, such as the use of platelet rich plasma (PRPs), growth factors, use of carriers or mesenchymal stem cells [2, 5–11]. All these bioaugmented techniques intend to improve the healing of the tendon-bone interface of the rotator cuff [2]. The enthesis or footprint is a complex structure formed by two very distinct tissues (tendon and bone) and is characterised by four distinct zones with a continuous transition of tissue material [2]. The histological aspect of this transition zone is altered in chronic and acute rotator tears affecting the biomechanical properties of the tendon. It is therefore important to have a reliable and reproducible experimental animal model simulating these changes in order to be allowed to evaluate the effectiveness of new techniques.

Our research group has developed, after several years of studies, a solid chronic injury model, which can also be adapted to create an acute injury model. The objective of this chapter is to describe our surgical technique used: anaesthesia of the animal, the surgical process, postoperative care and the euthanasia [12–14].

## 2. Material and method

### 2.1 Animal of experimentation

There are many animals that have been studied in order to elucidate which is the best model to study in rotator cuff pathology. Soslowsky et al. carried out a study in different animals in which they evaluated 34 essential variables to establish an optimal model for experimental study of the rotator cuff [15]. The rat was considered the most suitable due to its similar anatomy and shoulder mobility. In humans, the supraspinatus tendon passes under the arch formed by the coracoid, acromion and coracoacromial ligament, while in the rat, this arch is formed by the coracoid, clavicle and acromion, as well as the ligaments that connect these structures. Anatomically, therefore, it is the animal that most resembles humans. Furthermore, the repetitive movement that they can perform when running, digging or eating predisposes the supraspinatus tendon of the rat to suffer chronic injuries similar to those that occur in humans, when sliding under this arch, and all of these tell us that despite being a quadruped, the upper extremities are not only load-bearing joints but also maintain other interesting functions to establish the rat as an ideal experimental model for the study of rotator cuff pathology.

Furthermore, it is an animal that has previously been widely used as an experimental animal in multiple models [16–18]. In a recent meta-analysis described by Yang et al. in which they reviewed all the experimental models in which transporters have been used for rotator cuff pathology, they included a total of 74 studies; of which 35 are described in rats, 22 in rabbits, 10 in sheep and 7 in dogs. With this data, we see that the rat model is the most used in the scientific community for these purposes [5].

To all those anatomical and functional advantages described, we must add others such as their easy handling and a short gestation period and that they are animals with rapid growth and a short half-life. The laboratories are conditioned for use, they take up little space and, from an economic point of view, their daily maintenance is affordable compared to other species.

The age of the animal used in the experimental model is a factor that we consider important when creating an injury model. An essential variable in the success rate after a rotator cuff repair is the healing capacity. As we well know, vascularisation capacity is an important factor and it has been shown that it also decreases with advancing age. For our study, mature 8-month-old rats were used (the average life of the rat is 30 months), in which the reparative response is already considerably diminished, thus reproducing more exactly what happened in humans.

When designing a study, we must always bear in mind the 3 R's principle (replace, reduce and refine), in order to use the minimum number of animals possible to achieve our main objective. In the case of rotator cuff surgery, we should only perform surgery in one of the shoulders in order to not create a postsurgical functional disability.

### 2.2 Required materials

In order to carry out the surgeries, we require the following:

- Sterile gloves, sterile surgical field, alcohol: the surgery should be performed under maximum sterile conditions.

- Suture material:

    - Vicryl Rapide 4.0: absorbable suture, which does not usually generate a foreign body reaction in the tissues. It is the suture used to close the deltoid, the subcutaneous cellular tissue and the creation of an intradermal suture of the skin.

    - Prolene 6.0: non-absorbable monofilament synthetic suture. It is the suture used both to mark the tendon end in the supraspinatus section in the first surgical intervention of the chronic injury model and to perform the tendon repair according to the modified Mason-Allen technique, both for the acute injury model and chronicle. See **Figure 1**. There are several articles that describe the combination of the use of this suture with the modified Mason-Allen technique [19–23].

- Gauze, sterile swabs and haemostats: necessary to dry and clean the surgical field of possible bleeding during the intervention.

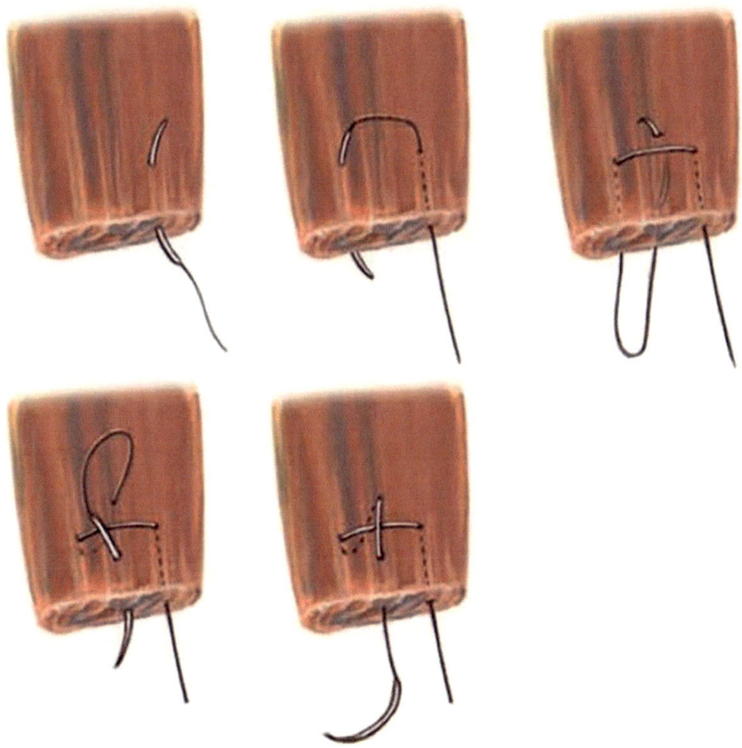

**Figure 1.**
*Modified Mason-Allen suture.*

- Microsurgery surgical instruments.
- Surgical microscope.
- Instrumentation necessary for local anaesthesia/orotracheal intubation.

## 2.3 Surgical technique

### 2.3.1 Acute rotator cuff tear

This injury model requires only one surgical intervention, in which the supraspinatus is sectioned, and its repair is immediately performed using suture.

#### 2.3.1.1 Anaesthesia

This single intervention requires a long surgical time, so it will be necessary to administer general anaesthesia to the animal, with subsequent orotracheal intubation.

To avoid the stress generated by the injection of a drug in an animal, oxygen and 3–4% isoflurane are first administered through an anaesthesia mask. Next, intraperitoneal ketamine 75 mg/kg and medetomidine 0.25 mg/kg are injected after which an orotracheal intubation with a 14G intravenous catheter is performed. Once the airway is controlled, we connected the animal to the respirator with mechanical ventilation by pressure of 12–14 cm $H_2O$ with an inspiration/expiration ratio of 1:2 and with a respiratory rate of 60 rpm. Anaesthetic maintenance will be performed with 2% isoflurane, inspired $O_2$ fraction of 0.2 and positive pressure at the end of expiration of 4 cm of $H_2O$.

The respirator we use is a Datex-Ohmeda pressure volume control. In addition, a dose of enrofloxacin 10 mg/kg is administered subcutaneously as prophylactic antibiotic therapy before beginning the surgical intervention.

| | Medication | Posology |
|---|---|---|
| General anaesthesia | Isoflurane 3–4% | Inhaled in anaesthetic induction |
| | Isoflurane 2% | Inhaled in anaesthetic maintenance |
| Local anaesthesia | Mepivacaine 2% | 1 mg/kg subcutaneous |
| Antibiotic prophylaxis | Enrofloxacin | 10 mg/kg subcutaneous |
| Preintubation | Ketamine | 75 mg/kg intraperitoneal |
| | Medetomidine | 0.25 mg/kg intraperitoneal |
| Reverse anaesthesia | Atipamezole | 0.1 mg/kg subcutaneous |
| Immediate postoperative pain | Thiopental | 40 mg/kg intraperitoneal |
| | Fentanyl | 0.3 mg/kg intraperitoneal |
| 3 days postoperative | Enrofloxacin | 10 mg/kg/24 h subcutaneous |
| | Meloxicam | 0.2 mg/kg/24 h subcutaneous |

**Table 1.**
*Anaesthetics and analgesics required for surgical procedures in rats.*

Once the intervention is finished, the gas is removed, leaving the animal alone with oxygen, and we administer atipamezole 0.1 mg/kg subcutaneously to reverse the anaesthesia (**Table 1**) [24–27].

#### *2.3.1.2 Rotator cuff tear and repair*

Once the animal has been anaesthetised, the anterior region of the shoulder should be shaved. The rat is put in a supine position with the head closest to the surgeon, with the paw in internal rotation strapped under its own body in order to more easily expose the humeral head. The rat should be placed on a thermal blanket for its well-being. Finally, the surgical field is painted with alcohol, and the rest should be covered with a sterile cloth.

The surgery should be performed with a microscopic view (microscope or magnifier). A skin incision of about 2 cm is made over the left shoulder with a scalpel over the superolateral part of the scapulohumeral joint. We dissect the subcutaneous cellular tissue and fat until the deltoid is exposed. An inverted T incision is made in the muscle, disinserting the proximal part of the anterior, lateral and posterior deltoid from the acromion and clavicle (**Figure 2**). Once completed, we visualise the supraspinatus tendon passing under the bony arch formed by the acromion, coracoid and clavicle, as well as the long head of the biceps (**Figures 3** and **4**).

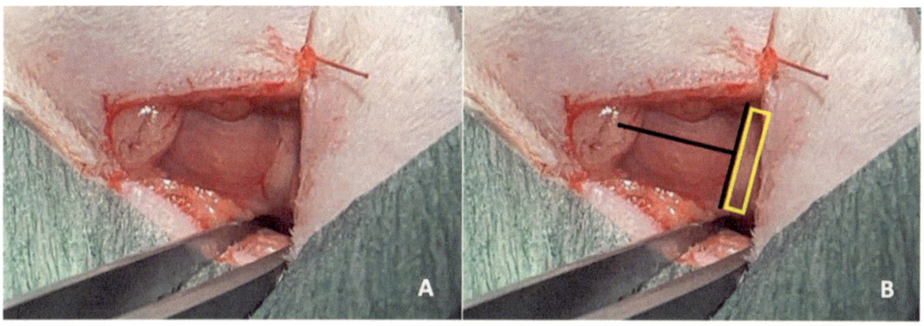


**Figure 2.**
*A: Approach of about 2 centimetres over the anterior scapulohumeral joint. Exposure of the deltoid muscle and clavicle. B: Same image on which the inverted T incision on the deltoid is shown in black. In yellow the clavicle.*

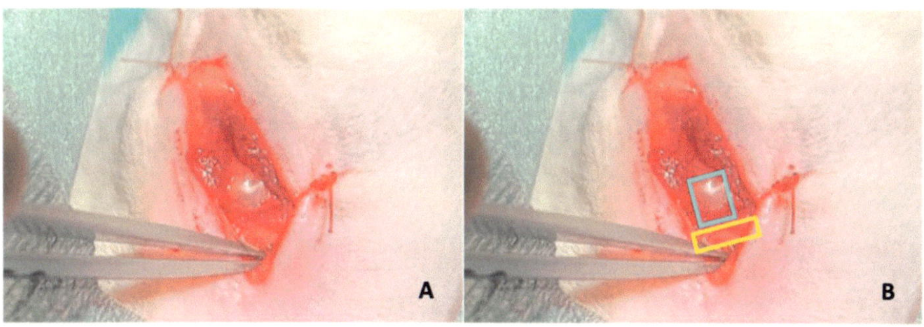


**Figure 3.**
*A: After the inverted T opening of the deltoid, the pearly white supraspinatus tendon is visualised inserting into the humeral head, appearing under the clavicle. B: Same image on which the supraspinatus is shown in blue and the clavicle in yellow.*

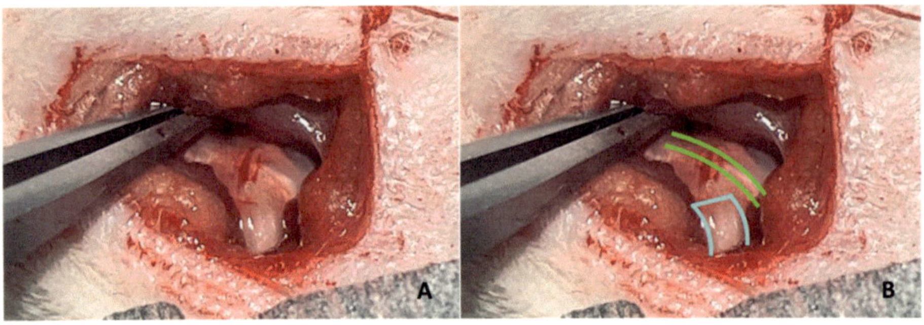


**Figure 4.**
*A: We retract the deltoid medially, and through transparency, we see the supraspinatus tendon and the long head of the biceps. B: Same image on which the supraspinatus is shown in blue and the long head of the biceps in green.*

The tendon should ten be delimited its anterior and posterior part, and its section should be as perpendicular as possible to its longitudinal axis and at a distance of approximately 4 mm from its insertion.

Once sectioned, tendon reparation should be performed in the same surgical act with a Mason-Allen–type suture modified with Prolene® 6.0, as described in **Figure 1**. **Figure 5** shows the suture in the animal.

For its fixation in the humerus, a transosseous tunnel is performed with a crab-type clamp in which the needle of the tendon suture is passed (using a subcutaneous needle) thus bringing the proximal end of the supraspinatus tendon closer to its greater tuberosity imprint (**Figures 6** and 7). Once the tendon is sutured, we proceed to inject or interpose that substance under study in our work to be carried out (**Figure 8**). If we inject liquid, we recommend previously closing the deltoid, so that the glenohumeral joint remains as tight as possible, to avoid leaks. If what is being applied is a carrier, we recommend being careful with the size, so as not to exceed too much beyond the thickness of the supraspinatus tendon.

For the closure of the surgical wound a Vicryl Rapide® 4.0 should be used to suture muscle, subcutaneous tissue and skin (**Figure 9**).

### *2.3.1.3 Postoperative care*

During the postoperative period, enrofloxacin 10 mg/kg/24 h subcutaneously and meloxicam 0.2 mg/kg/24 h subcutaneously were injected for 3 days [24–27].

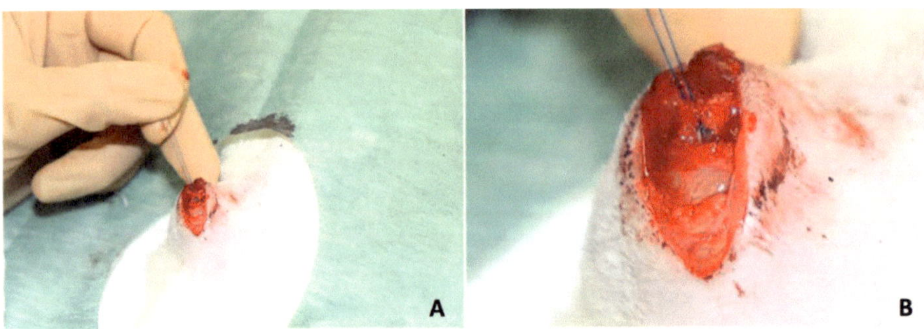


**Figure 5.**
*A: Modified Mason-Allen–type suture. B: enlarged image of A, where the suture performed is better visualised.*

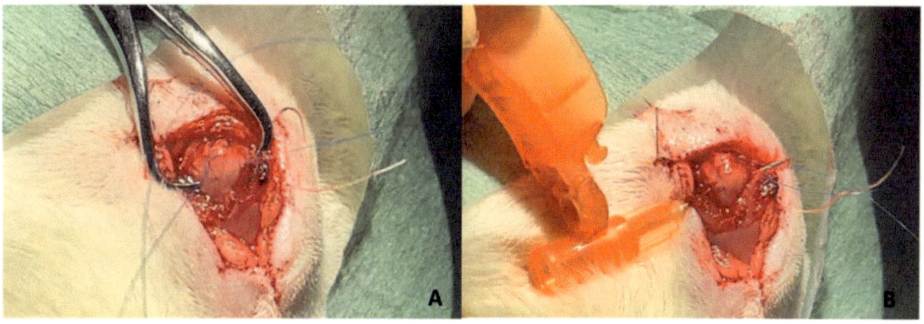


**Figure 6.**
*A: Creation of the transosseous tunnel with a crab-type clamp. B: Channelling with a subcutaneous needle to later pass the end of the needle of the suture made to the tendon.*

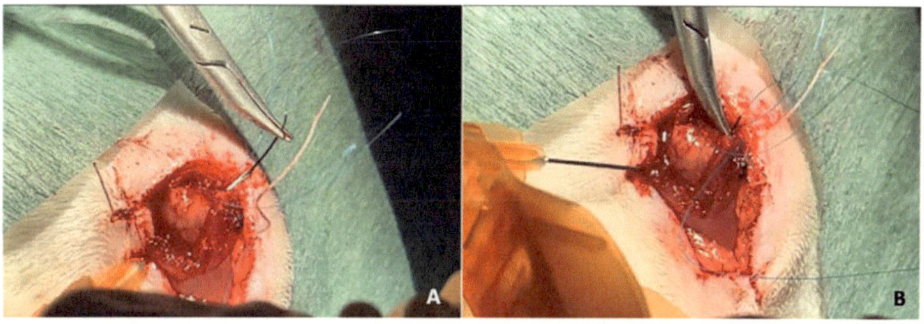


**Figure 7.**
*A: Beginning of the assisted suture needle passage of the subcutaneous needle. B: Needle passing the humeral transosseous tunnel.*

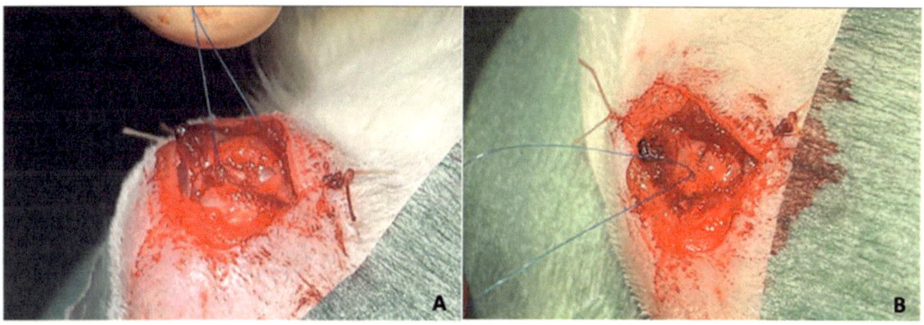


**Figure 8.**
*A: Both ends of Prolene 6.0. B: Prolene sutured, placing the supraspinatus tendon on the humeral imprint.*

## *2.3.2 Chronic rotator cuff tear*

This injury model requires two interventions, the first to perform the tendon section and the second to perform the repair of the supraspinatus 4 weeks after the injury, which is when histological changes similar to those seen in chronic rotator cuff lesions in human beings are obtained [12–14].

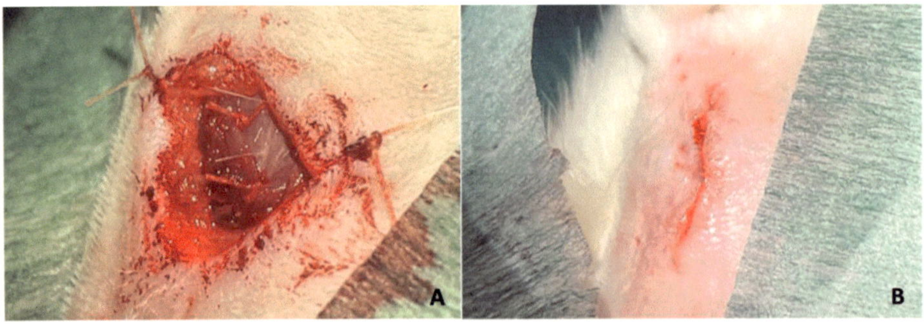


**Figure 9.**
*A: Closure of the muscular plane. B: Closure of the skin with intradermal suture without leaving ends so that the rats cannot bite themselves.*

### *2.3.2.1 Anaesthesia*

To perform the first intervention, or surgery on the injury, we perform it under sedation with isoflurane without the need for orotracheal intubation of the animal, given that it is a shorter intervention, in order to reduce anaesthetic risks and therapeutic aggressiveness.

Firstly, oxygen and 3–4% isoflurane are supplied to the animal for anaesthetic induction and subsequently the gas is left at 2% for maintenance. The intervention is performed with local anaesthetic, injecting 0.6 cc of 2% mepivacaine into the skin and subcutaneous tissue (for a rat weighing approximately 600 g) [24–27].

Once the intervention is finished, the gas is removed and the animal is left alone with oxygen, until it eliminates the inhaled gas itself and gradually comes to its senses.

To perform the second intervention, or the repair surgery, we will carry out the same procedure explained in 2.3.1.1.

### *2.3.2.2 Rotator cuff tear*

Once the animal has been anaesthetised, the anterior region of the shoulder should be shaved. The rat is put in a supine position with the head closest to the surgeon, with the paw in internal rotation strapped under its own body in order to more easily expose the humeral head. The rat should be placed on a thermal blanket for its well-being. Finally, the surgical field is painted with alcohol, and the rest should be covered with a sterile cloth.

When performing the surgery we recommend using a microscopic view (microscope or magnifier). A skin incision of about 2 cm is made over the left shoulder with a scalpel over the superolateral part of the scapulohumeral joint. We dissect the subcutaneous cellular tissue and fat until the deltoid is exposed. An inverted T incision is made in the muscle, disinserting the proximal part of the anterior, lateral and posterior deltoid from the acromion and clavicle (**Figure 2**). Once completed, we visualise the supraspinatus tendon passing under the bony arch formed by the acromion, coracoid and clavicle, as well as the long head of the biceps (**Figures 3** and **4**).

We clearly delimit the tendon in its anterior and posterior part, and its section should be as perpendicular as possible to its longitudinal axis and at a distance of approximately 4 mm from its insertion.

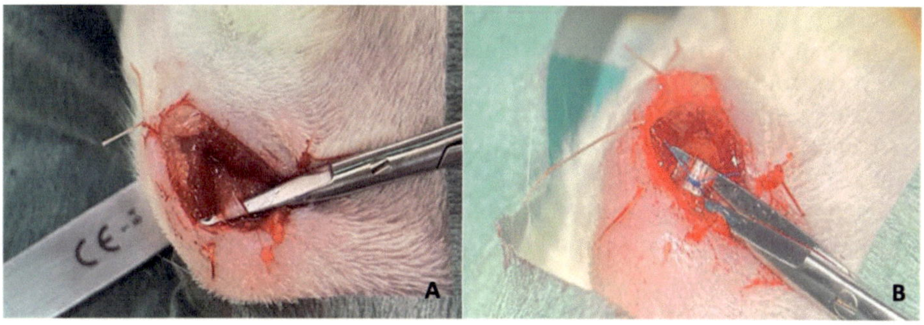


**Figure 10.**
*A: Delimitation of the supraspinatus tendon. B: Marking with Prolene 6.0 of the tendon after its section.*

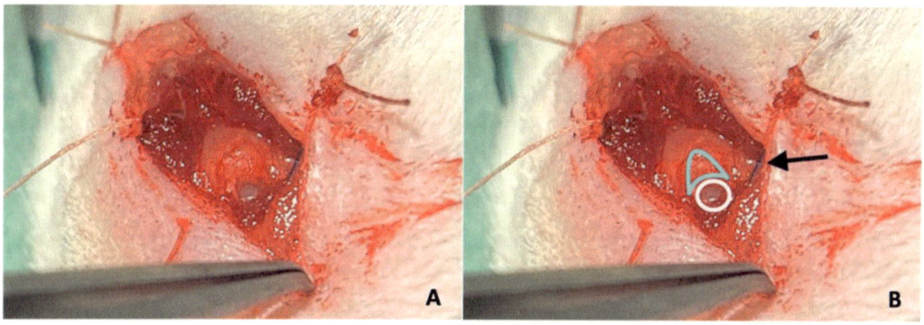


**Figure 11.**
*Once the tendon has been sectioned, we see a stump of about 4 mm inserted on the humeral tuberosity and the cartilage of the humeral head exposed. B: Same image on which the disinserted supraspinatus stump is shown in blue and the humeral cartilage in white. The black arrow points to the Prolene marking of the medially retracted supraspinatus tendon.*

We clearly delimit the tendon in its anterior and posterior part and mark it in its proximal region with a Prolene® 6.0 to later make the section of the tendon as perpendicular as possible to its long axis and at a distance of approximately 4 mm from its insertion. We see how the tendon retracts proximally, allowing us to see the exposed cartilage of the humeral head (**Figures 10** and **11**).

During the surgical technique, special care will be taken to keep the rest of the tendons of the cuff and the structures of the scapulohumeral joint intact. Finally, muscle, subcutaneous tissue and skin will be sutured with Vicryl Rapide® 4.0 (**Figure 9**).

### *2.3.2.3 Rotator cuff repair*

Once 4 weeks have passed after the first intervention, we perform the repair surgery. Once the animal has been anaesthetised, the rat is prepared as described previously. The same incisions are performed in the same way as was done in the first intervention. After the deltoid incision, a more fibrous tissue appears compared to the first intervention and we localise one end of the tendon in the region of the greater tuberosity, which must be debrided. Once this is done, the retracted proximal end should be found using the mark of Prolene® from the first surgery. Once located and dissected of fibrosis, we will perform the modified Mason-Allen–type suture with Prolene® 6.0 (**Figure 5**).

For its fixation in the humerus, a transosseous tunnel is performed with a crab-type clamp in which the needle of the tendon suture is passed (using a subcutaneous needle), thus bringing the proximal end of the supraspinatus tendon closer to its greater tuberosity imprint (**Figures 6** and 7). Once the tendon is sutured, we proceed to inject or interpose that substance under study in our work to be carried out (**Figure 8**). If injecting liquid, we recommend previously closing the deltoid, so that the glenohumeral joint remains as tight as possible, to avoid leaks. If what is being applied is a carrier, we recommend being careful with the size, so as not to exceed too much beyond the thickness of the supraspinatus tendon.

Finally, the muscle, subcutaneous tissue and skin will be sutured with Vicryl Rapide® 4.0. We recommend using intradermal sutures without leaving visible ends, so that rodents do not bite the wounds (**Figure 9**).

#### *2.3.2.4 Postoperative care*

During the postoperative period, enrofloxacin 10 mg/kg/24 h subcutaneously and meloxicam 0.2 mg/kg/24 h subcutaneously were injected for 3 days [24–27].

### *2.3.3 Euthanasia*

Euthanasia is performed 4 months after the tendon repair in both acute and chronic model, in order to permit tendon remodelling, which should give more accurate results [28].

The phases of tendon remodelling are summarised in three periods:

1. The inflammation phase, which occurs in the first week after the intervention. Platelet aggregation, fibrin deposition and the release of different cytokines are the protagonists in this phase.

2. The repair phase, which occurs during the first 8 weeks, in which macrophage aggregates, and fibroblast proliferation and type III collagen deposits predominate.

3. The remodelling phase, which can last from a few months to a year. During this phase, type III collagen is converted to type I collagen, and that is why it is important to make way for this phase prior to euthanasia of the animal.

For euthanasia, 4% isoflurane is applied through an anaesthesia mask until the animal is sedated. Subsequently, the animal is placed in a hermetically closed cabin with a $CO_2$ atmosphere.

## 2.4 Biological implantation and analysis of results

There are many possibilities for biological implantation, which have been studied over the years in these tendon injury models. In our injury model, all types of therapies can be provided, both in suspension and using scaffolds as transporters (such as growth factors, mesenchymal stem cells derived from bone marrow or mesenchymal stem cells derived from lipoaspirate).

In the literature, there are multiple studies that try to establish which biological implementation would permit an improved rotator cuff tendon healing.

PRPs are one of the most common biological products used in many human studies. They have the advantage of being very affordable; however, the wide variety of studies that exist makes it difficult to obtain a standardised process for their use [29–32].

Growth factors can promote tendon healing, but their cost and accelerated metabolism make them difficult to work with in animal research. They are therefore often used in conjunction with mesenchymal stem cells and scaffolds [6].

As for the use of mesenchymal stem cells, the most widely used over the years have been those derived from bone marrow stem cells (BMSCs) (**Table 2**) [11, 12, 17, 20, 33–41]; however, the great accessibility and availability of large quantities of cells to create cell therapy drugs have led to an increased interest in adipose derived stem cells (ADSCs) in recent years (**Table 3**) [13, 21, 42–48].

As to the evaluation of the results, we recommend at least a histological analysis of the tendon that objectifies the tendon healing obtained with the treatment and, on the other hand, a biomechanical study that assesses the resistance and elasticity of the tendon [13].

There are other interesting studies that could be associated, such as immunohistochemistry techniques or techniques that assess the biomechanics of the tendon in the animal in vivo [50].

## 3. Discussion

Models of acute injury repair in rats are widely described. In **Tables 2** and **3**, it is reflected that most studies are performed using an acute model; however, the healing after acute injury repair differs from healing after chronic repair [51, 52]. It is therefore important to consider that if the aim is to improve tendon healing in patients with chronic injuries then the study should based on an experimental chronic injury model.

Buchmann et al. performed a tendon section in 45 12-week-old Sprague-Dawley rats to evaluate them macroscopically and histologically at 3, 6 and 9 weeks (15 rats in each group), simply to see the histological evolution of the lesion in acute and chronic injuries. They observed an osteotendinous gap in 60% of the animals at 3 weeks, which decreased in the other 2 groups. At the histological level, they observed that muscle degeneration had decreased to normal values from the ninth postoperative week onwards. Meanwhile the tendon structure showed values similar to those of a normal tendon at 9 weeks compared to 3 and 6 weeks. Therefore, it should be noted that chronic injury in rats is maintained until at least the third week, but in young animals such as those used in the study, they end up self-regenerating at the ninth week after tendon sectioning [51].

Following a similar line, our working group opted to repair the lesion 4 weeks after injury, considering at this point that the lesion is chronic. After repair at 1 month, the lesion was evaluated at different time points (1, 2 and 3 months) to analyse the histological progression of the lesion. The results validated our model of chronic injury [12, 28, 53, 54]. Histologically, tendon retraction with degeneration and fibrosis of the edges and an osteotendinous gap were observed, while biomechanically, less strength and more stiffness were obtained compared

| | Animal | Injury model | Type of cells | Method of delivery/ Scaffold | Time of observation | Results |
|---|---|---|---|---|---|---|
| Gulotta et al. [33] | Rat | Acute repair of the supraspinatus tendon injury | BMSCs | Fibrin carrier | 2 and 4 weeks | No improvement of structure, composition and strength in the tendon junction area. |
| Gulotta et al. [34] | Rat | Acute repair of the supraspinatus tendon injury | BMSCs with MT1 – MMP scaffold | Fibrin carrier | 2 and 4 weeks | Presence of fibrocartilage in the lesion with improved biomechanical strength |
| Gulotta et al. [17, 35] | Rat | Acute repair of the supraspinatus tendon injury | BMSC with 'scleraxis' scaffold | Fibrin carrier | 4 and 8 weeks | Improved histological appearance, with more fibrocartilage and increased biomechanical strength |
| Gulotta et al. [17, 35] | Rat | Acute repair of the supraspinatus tendon injury | BMSC with BMP-13 scaffold | Fibrin carrier | 2 and 4 weeks | No difference between histological appearance and biomechanical repair strength |
| Kida et al. [11] | Rat | Acute repair of the supraspinatus tendon injury | BMSCs | Designed cells | 2, 4 and 8 weeks | BMSCs were infiltrated into the repaired tendon and improved biomechanical properties |
| Tornero-Esteban et al. [12] | Rat | Chronic repair of the supraspinatus tendon injury | BMSCs | Collagen carrier | 4, 8 and 12 weeks | Increase in peak load strength at 3 months compared to suture only group |
| Degen et al. [36] | Rat | Acute repair of the supraspinatus tendon injury | BMSCs | Fibrin carrier | 2 and 4 weeks | Improvement of initial histological appearance and biomechanical strength |
| Omi et al. [37] | Rat | Acute repair of the supraspinatus tendon injury | BMSCs | Multilayer xenograft tendon | 6 weeks | Solid repair with increased biomechanical forces from week 6 onwards |
| Peach et al. [20] | Rat | Acute repair of the supraspinatus tendon injury | BMSCs | PCL/ PNEAmPh fiber | 6 and 12 weeks | Increased strength and improved morphology of the structure at 6 and 12 weeks |
| Zong et al. [38] | Rat | Acute repair of the supraspinatus tendon injury | BMSCs | Fibrin carrier | 2 and 4 weeks | BMSCs have a positive effect on the formation of fibrocartilage |
| Thangarajah et al. [39] | Rat | Chronic repair of the supraspinatus tendon injury | BMSCs | Fibrin carrier | 6 weeks | Improves rotator cuff healing and restores bone mineral density |

| | Animal | Injury model | Type of cells | Method of delivery/ Scaffold | Time of observation | Results |
|---|---|---|---|---|---|---|
| Gülecyüz et al. [40] | Rat | Chronic repair of the supraspinatus tendon injury | BMSCs | | 8 weeks | The BMSC group showed an initially significantly higher muscle mass, but no difference between groups at week 8 |
| Han et al. [41] | Rat | Acute repair of the supraspinatus tendon injury | BMSCs | | 4 and 8 weeks | Improves the biomechanical properties of the tendon |

**Table 2.**
*Therapeutic effects of BMSCs on rotator cuff injuries in rats.*

to healthy shoulders. Although Buchmann et al. [51] indicated that chronic injury was best reproduced in the third week after section, they were not able to observe histological differences between the third and sixth week; therefore, we opted to follow the chronic injury model previously used by our working group, considering the injury chronic after 4 weeks [12, 14, 28, 53, 54].

Confirming these data, in two previous studies also performed by our group, there was no evidence of spontaneous regeneration 4 weeks after section in any animal. All the cases had an osteotendinous space where the supraspinatus tendon should have been, so there was no natural tendon recovery at the time of reintervention [49, 53].

The modified Mason-Allen–type suture has been widely used for this type of study, and different authors such as Gerber or Derwin carried out biomechanical studies and strength studies on this type of suture with good results [12, 14, 16, 22, 54]. Gerber compared, from a biomechanical point of view, different sutures made in ovine infraspinatus tendons, obtaining a maximum load of up to 359 Newton with 2 transosseous modified Mason-Allen stitches [19, 55]. Derwin et al. found that despite having only two-thirds the strength of a Krakow-type suture, a modified Mason-Allen suture passes less often through the tendon, providing a good balance between strength and tendon preservation [56], a very important point to consider given the size of a rat's supraspinatus tendon.

The choice of suture material is crucial for the stability of the repair. Braided polyester (e.g. Ethibond®) is a very stiff material with a high peak load, while absorbable braided sutures such as Vicryl® have similar mechanical properties in vitro, but once implanted in the subcutaneous cellular tissue or muscle, the strength decreases by 50%. Absorbable monofilaments also lose 50% of their initial strength after implantation, in this case after 3–5 weeks. This makes absorbable sutures, both braided and monofilament, unsuitable for rotator cuff repair as at least 6 weeks are needed to ensure biological fixation of the rotator cuff. Prolene® has a low coefficient of friction and a high elastic recovery capacity while creating a minimal inflammatory reaction in the tissues and has been described as an augmentation patch with good results [57]. Given the size of the supraspinatus tendon in rats, we chose to use a 6/0 nonabsorbable monofilament suture of the Prolene® type.

According to the literature reviewed, other studies such as Peach et al. also used this combination of lesion model with a modified Mason-Allen suture with a 6/0 Prolene® [20]. However, Kaizawa et al. used the modified Mason-Allen suture but

| | Animal | Injury model | Type of cells | Method of delivery/ scaffold | Time of observation | Results |
|---|---|---|---|---|---|---|
| Valencia Mora et al. [42] | Rat | Chronic repair of the supraspinatus tendon | ADSCs | Collagen carrier | 2 and 4 weeks | Less inflammation. No improvement of tendon biomechanics |
| Barco et al. [43] | Rat | Acute repair of the supraspinatus tendon injury | ADSCs | Fibrin scaffold | 4 and 8 weeks | Less presence of neutrophils and more presence of plasma cells without improving the histological appearance and biomechanical strength of the tendon |
| Chen et al. [44] | Rat | Acute repair of the supraspinatus tendon injury | ADSCs | | 1 week | Decrease in inflammation, improvement in histological appearance and load to failure after 7 days, no difference thereafter |
| Lipner et al. [45] | Rat | Acute repair of the supraspinatus tendon injury | ADSCs/ ADSCs with BMP-12 | PLGA nanofibres with Poly lactic co-glycolic acid fibrin hydrogel | 2, 4 and 8 weeks | Decreased biomechanical properties of the tendon at the repair site |
| Rothrauff et al. [46] | Rat | Acute repair of the supraspinatus and infraspinatus/chronic repair and intramuscular injection of botulinum toxin A | ADSCs | GelMA/fibrin hydrogel | 4 weeks | Improved bone mineral density of the proximal humerus |
| Kaizawa et al. [21, 23] | Rat | Chronic repair of the supraspinatus tendon injury | ADSCs | Human tendon hydrogel | 8 weeks | Improved biomechanical properties of the tendon and fibrocartilage area, but no improvement of the bone interface area |
| Shin et al. [47] | Rat | Chronic repair of the supraspinatus tendon injury | ADSCs | Designed cells | 2 and 4 weeks | Increased fibrocartilage area, with improved density volume and biomechanical properties of the tendon |
| Choi et al. [48] | Rat | Chronic repair of the supraspinatus tendon injury | ADSCs | Sheets of cells | 2 weeks | Stem cell sheets enhanced regeneration of the tendon-to-bone junction when interpositioned at the tendon-to-bone interface |
| Ponz-Lueza et al. [49] | Rat | Chronic repair of the supraspinatus tendon injury | ADSCs | | 4 weeks | No histological or biomechanical improvement of the tendon |

**Table 3.**
*Therapeutic effects of ADSCs on rotator cuff injuries in rats.*

with a 5/0 Prolene® [21, 23] in both their studies according to the technique previously described by Thomopoulos et al. [22].

## 4. Conclusions

After many years of study and different research works, we consider that our model of chronic injury in rats is a reproducible model and that it manages to simulate chronic injuries of the rotator cuff in humans in satisfactory terms for the development of clinical research in this type of injuries.

Furthermore, it presents great variability since, based on our injury model, and depending on what we want to investigate, certain items can be varied to achieve the established objectives, such as adapting the model to reproduce an acute injury as described in the first phase of the article or providing different treatments to the model, varying from the use of transporters to the use of mesenchymal stem cells in suspension.

## Author details

Virginia Ponz-Lueza[1]*, Yaiza Lopiz[1,2,3], Camilla Arvinius[1], Cruz Rodriguez-Bobada[2] and Fernando Marco[1,2,3]

1 Shoulder and Elbow Unit, Department of Traumatology and Orthopaedic Surgery, Clínico San Carlos Hospital, Madrid, Spain

2 Unit of Experimental Surgery, Hospital Clínico San Carlos, Madrid, Spain

3 Department of Surgery, Complutense University, Madrid, Spain

*Address all correspondence to: virginiaponz@gmail.com

## References

[1] Yamaguchi K, Ditsios K, Middleton WD, Hildebolt CF, Galatz LM, Teefey SA. The demographic and morphological features of rotator cuff disease. A comparison of asymptomatic and symp-tomatic shoulders. The Journal of Bone and Joint Surgery. American Volume. 2006;**88**:1699-1704. DOI: 10.2106/JBJS.E.00835

[2] Zhang C, Wu J, Li X, Wang Z, Lu WW, Wong TM. Current biological strategies to enhance surgical treatment for Rota-tor cuff repair. Frontiers in Bioengineering and Biotechnology. 2021;**9**:657584. DOI: 10.3389/fbioc.2021.657584

[3] Lopiz Morales Y, García Fernández C, Vega ML, Marco F. Evalua-ción clínico-ecográfica de la reparación quirúrgica de las roturasdel manguito rotador. Trauma Fund MAPFRE. 2010;**21**:91-96

[4] Lafosse L, Brozska R, Toussaint B, Gobezie R. The out-come and structural integrity of arthroscopic rotatorcuff repair with use of the double-row suture anchortechnique. The Journal of Bone and Joint Surgery. American Volume. 2007;**89**:1533-1541. DOI: 10.2106/JBJS.F.00305

[5] Yang J, Kang Y, Zhao W, Jiang J, Jiang Y, Zhao B, et al. Evaluation of patches for rotator cuff repair: A systematic review and meta-analysis based on animal studies. Bioactive Materials. 2022;**10**:474-491

[6] Gulotta LV, Rodeo SA. Growth factors for rotator cuff repair. Clinics in Sports Medicine. 2009;**28**(1):13-23

[7] Molloy T, Wang Y, Murrell G. The roles of growth factors in tendon and ligament healing. Sports Medicine. 2003;**33**(5):381-394

[8] McCormack RA, Shreve M, Strauss EJ. Biologic augmentation in rotator cuff repair—should we do it, who should get it, and has it worked? Bulletin/Hospital for Joint Diseases (2013). 2014;**72**(1):89-96

[9] Heo JS, Choi Y, Kim HS, Kim HO. Comparison of molecular profiles of human mesenchymal stem cells derived from bone marrow, umbilical cord blood, placenta and adipose tissue. International Journal of Molecular Medicine. 2016;**37**(1):115-125

[10] Liu Q, Yu Y, Reisdorf RL, Qi J, Lu CK, Berglund LJ, et al. Engineered tendonfibrocartilagebone composite and bone marrow-derived mesenchymal stem cell sheet augmentation promotes rotator cuff healing in a non-weight-bearing canine model. Biomaterials. 2019;**192**:189-198

[11] Kida Y, Morihara T, Matsuda K, Kajikawa Y, Tachiiri H, Iwata Y, et al. Bone marrowderived cells from the footprint infiltrate into the repaired rotator cuff. Journal of Shoulder and Elbow Surgery. 2013;**22**(2):197-205

[12] Tornero-Esteban P, Hoyas JA, Villafuertes E, Rodríguez-Bobada C, López-Gordillo Y, Rojo FJ, et al. Efficacy of supraspinatus tendon repair using mesenchymal stem cells along with a collagen I scaffold. Journal of Orthopaedic Surgery and Research. 2015;**10**:124. DOI: 10.1186/s13018-015-0269-6

[13] Ponz-Lueza V, Lopiz Y, Rodríguez-Bobada C, Tornero-Esteban P, Arvinius C, García-Fernández C, et al. Efficacy of transplantation of lipoaspired mesenchymal stem cells in the treatment of chronic rotator

cuff tears. Experimental model in rats. Revista Española de Cirugía Ortopédica y Traumatología. English, Spanish. Jan-Feb 2024;**68**(1):9-17. DOI: 10.1016/j.recot.2023.05.006. Epub 2023 May 23. PMID: 37230410

[14] Arvinius C, Civantos A, Rodríguez-Bobada C, Rojo FJ, Pérez-Gallego D, Lopiz Y, et al. Enhancement of in vivo supraspinatus tendon-to-bone healing with an alginate-chitin scaffold and rhBMP-2. Injury. 2021;**52**(1):78-84. DOI: 10.1016/j.injury.2020.11.019. Epub 2020 Nov 9

[15] Soslowsky LJ, Carpenter JE, DeBano CM, Banerji I, Moalli MR. Development and use of an animal model for investigations on rotator cuff disease. Journal of Shoulder and Elbow Surgery. 1996;**5**(5):383-392

[16] Galatz LM, Sandell LJ, Rothermich SY, Das R, Mastny A, Havlioglu N, et al. Characteristics of the rat supraspinatus tendon during tendon-to-bone healing after acute injury. Journal of Orthopaedic Research. 2006;**24**(3):541-550

[17] Gulotta LV, Kovacevic D, Packer JD, Deng XH, Rodeo SA. Bone marrow-derived mesenchymal stem cells transduced with scleraxis improve rotator cuff healing in a rat model. The American Journal of Sports Medicine. 2011;**39**(6):1282-1289. DOI: 10.1177/0363546510395485. Epub 2011 Feb 18

[18] Oztermeli A, Karaca S, Yucel I, Midi A, Sen EI, Ozturk BY. The effect of erythropoietin on rat rotator cuff repair model: An experimental study. Journal of Orthopaedic Surgery (Hong Kong). 2019;**27**(2):2309499019856389. DOI: 10.1177/2309499019856389

[19] Gerber C, Schneeberger AG, Beck M, Schlegel U. Mechanical strength of repairs of the rotator cuff. Journal of Bone and Joint Surgery. British Volume (London). 1994;**76**(3):371-380

[20] Peach MS, Ramos DM, James R, Morozowich NL, Mazzocca AD, Doty SB, et al. Engineered stem cell niche matrices for rotator cuff tendon regenerative engineering. PLoS One. 2017;**12**(4):e0174789

[21] Kaizawa Y, Franklin A, Leyden J, Behn AW, Tulu US, Sotelo Leon D, et al. Augmentation of chronic rotator cuff healing using adipose-derived stem cellseeded human tendon-derived hydrogel. Journal of Orthopaedic Research. 2019;**37**(4):877-886

[22] Thomopoulos S, Hattersley G, Rosen V, Mertens M, Galatz L, Williams GR, et al. The localized expression of extracellular matrix components in healing tendon insertion sites: An in situ hybridization study. Journal of Orthopaedic Research. 2002;**20**(3):454-463

[23] Kaizawa Y, Leyden J, Behn AW, Tulu US, Franklin A, Wang Z, et al. Human tendon-derived collagen hydrogel significantly improves biomechanical properties of the tendon-bone interface in a chronic rotator cuff injury model. The Journal of Hand Surgery. 2019;**44**(10):899.e1-899.e11

[24] Burnside WM, Flecknell PA, Cameron AI, Thomas AA. A comparison of medetomidine and its active enantiomer dexmedetomidine when administered with ketamine in mice. BMC Veterinary Research. 2013;**9**:48. DOI: 10.1186/1746-6148-9-48

[25] Flecknell P. Laboratory Animal Anaesthesia and Analgesia. 5th ed. Feb 2023. ISBN: 9780128182697

[26] Rodent analgesia: Assessment and therapeutics Paul Flecknell Institute of

Neuroscience, Newcastle University, Newcastle upon Tyne NE24HH, UK. The Veterinary Journal. 2018;**232**:70-77

[27] Flecknell P, Lofgren JLS, Dyson MC, Marini RR, Swindle MM, Wilson RP. Chapter 24 - Preanesthesia, anesthesia, analgesia, and euthanasia. In: Laboratory Animal Medicine. 3rd ed. Newcastle: American College of Laboratory Animal Medicine; 2015. pp. 1135-1200

[28] Rodriguez-Bobada C. Desarrollo de un modelo experimental de lesión-reparación del manguito rotador en rata para evaluación de procedimientos en clínica humana. [Tesis Doctoral]. Universidad Complutense de Madrid; 2013

[29] Randelli PS, Stoppani CA, Santarsiero G, Nocerino E, Menon A. Platelet-rich plasma in arthroscopic rotator cuff repair: Clinical and radiological results of a prospective randomized controlled trial study at 10-year follow-up. Arthroscopy. 2022;**38**(1):51-61

[30] Wu Y, Wang R, Meng Q. Clinical effect of PRP with DPASB on full-thickness rotator cuff tears and its role in VAS, SST, and constant scores of patients. Clinical Laboratory. 1 Aug 2021;**67**(8). DOI: 10.7754/Clin.Lab.2020.201031. PMID: 34383416

[31] Snow M, Hussain F, Pagkalos J, Kowalski T, Green M, Massoud S, et al. The effect of delayed injection of leukocyte-rich platelet-rich plasma following rotator cuff repair on patient function: A randomized double-blind controlled trial. Arthroscopy. 2020;**36**(3):648-657

[32] Malavolta EA, Gracitelli MEC, Assuncao JH, Ferreira Neto AA, Bordalo-Rodrigues M, de Camargo OP. Clinical and structural evaluations of rotator cuff repair with and without added platelet-rich plasma at 5-year follow-up: A prospective randomized study. The American Journal of Sports Medicine. 2018;**46**(13):3134-3141

[33] Gulotta LV, Kovacevic D, Ehteshami JR, Dagher E, Packer JD, Rodeo SA. Application of bone marrow-derived mesenchymal stem cells in a rotator cuff repair model. The American Journal of Sports Medicine. 2009;**37**(11):2126-2133

[34] Gulotta LV, Kovacevic D, Montgomery S, Ehteshami JR, Packer JD, Rodeo SA. Stem cells genetically modified with the developmental gene MT1-MMP improve regeneration of the supraspinatus tendon-to-bone insertion site. The American Journal of Sports Medicine. 2010;**38**(7):1429-1437

[35] Gulotta LV, Kovacevic D, Packer JD, Ehteshami JR, Rodeo SA. Adenoviral-mediated gene transfer of human bone morphogenetic protein-13 does not improve rotator cuff healing in a rat model. The American Journal of Sports Medicine. 2011;**39**(1):180-187

[36] Degen RM, Carbone A, Carballo C, Zong J, Chen T, Lebaschi A, et al. The effect of purified human bone marrow-derived mesenchymal stem cells on rotator cuff tendon healing in an athymic rat. Arthroscopy. 2016;**32**(12):2435-2443

[37] Omi R, Gingery A, Steinmann SP, Amadio PC, An KN, Zhao C. Rotator cuff repair augmentation in a rat model that combines a multilayer xenograft tendon scaffold with bone marrow stromal cells. Journal of Shoulder and Elbow Surgery. 2016;**25**(3):469-477

[38] Zong JC, Mosca MJ, Degen RM, Lebaschi A, Carballo C, Carbone A, et al. Involvement of Indian hedgehog

signaling in mesenchymal stem cell-augmented rotator cuff tendon repair in an athymic rat model. Journal of Shoulder and Elbow Surgery. 2017;**26**(4):580-588

[39] Thangarajah T, Sanghani-Kerai A, Henshaw F, Lambert SM, Pendegrass CJ, Blunn GW. Application of a demineralized cortical bone matrix and bone marrow-derived mesenchymal stem cells in a model of chronic rotator cuff degeneration. The American Journal of Sports Medicine. 2018;**46**(1):98-108

[40] Gulecyuz MF, Macha K, Pietschmann MF, Ficklscherer A, Sievers B, Rossbach BP, et al. Allogenic myocytes and mesenchymal stem cells partially improve fatty rotator cuff degeneration in a rat model. Stem Cell Reviews and Reports. 2018;**14**(6):847-859

[41] Han L, Fang WL, Jin B, Xu SC, Zheng X, Hu YG. Enhancement of tendon-bone healing after rotator cuff injuries using combined therapy with mesenchymal stem cells and platelet rich plasma. European Review for Medical and Pharmacological Sciences. 2019;**23**(20):9075-9084

[42] Valencia Mora M, Antuna S, Garcia Arranz M, Carrascal MT, Barco R. Application of adipose tissue-derived stem cells in a rat rotator cuff repair model. Injury. 2014;**45**(Suppl. 4):S22-S27

[43] Barco R, Encinas C, Valencia M, Carrascal MT, Garcia-Arranz M, Antuna S. Use of adipose-derived stem cells in an experimental rotator cuff fracture animal model. Revista Española de Cirugía Ortopédica y Traumatología. 2015;**59**(1):3-8

[44] Chen HS, Su YT, Chan TM, Su YJ, Syu WS, Harn HJ, et al. Human adipose-derived stem cells accelerate the restoration of tensile strength of tendon and alleviate the progression of rotator cuff injury in a rat model. Cell Transplantation. 2015;**24**(3):509-520

[45] Lipner J, Shen H, Cavinatto L, Liu W, Havlioglu N, Xia Y, et al. In vivo evaluation of adipose-derived stromal cells delivered with a nanofiber scaffold for tendon-to-bone repair. Tissue Engineering. Part A. 2015;**21**(21-22):2766-2774

[46] Rothrauff BB, Smith CA, Ferrer GA, Novaretti JV, Pauyo T, Chao T, et al. The effect of adipose-derived stem cells on enthesis healing after repair of acute and chronic massive rotator cuff tears in rats. Journal of Shoulder and Elbow Surgery. 2019;**28**(4):654-664

[47] Shin MJ, Shim IK, Kim DM, Choi JH, Lee YN, Jeon IH, et al. Engineered cell sheets for the effective delivery of adipose-derived stem cells for tendon-to-bone healing. The American Journal of Sports Medicine. 2020;**47**(13):3347-3358

[48] Choi JH, Shim IK, Shin MJ, Lee YN, Koh KH. Stem cell sheet interpositioned between the tendon and bone would be better for healing than stem cell sheet overlaid above the tendon-to-bone junction in rotator cuff repair of rats. PLoS One. 2022;**17**(3):e0266030. DOI: 10.1371/journal.pone.0266030

[49] Ponz-Lueza V. Seguridad y eficacia del trasplante de células madre mesenquimales de lipoaspirado en el tratamiento de lesiones crónicas del manguito rotador. Modelo experimental en ratones y ratas. [Tesis Doctoral]. Universidad Complutense de Madrid; 2024

[50] Liu Y, Fu SC, Yao SY, Yung PS. Skilled reaching test for shoulder function assessment in a rat model of rotator cuff tear: A pilot study. BMC Musculoskeletal

Disorders. 2024;**25**(1):506. DOI: 10.1186/s12891-024-07624-6

[51] Buchmann S, Walz L, Sandmann GH, Hoppe H, Beitzel K, Wexel G, et al. Rotator cuff changes in a full thickness tear rat model: Verification of the optimal time interval until reconstruction for comparison to the healing process of chronic lesions in humans. Archives of Orthopaedic and Trauma Surgery. 2011;**131**(3):429-435

[52] Killian ML, Cavinatto LM, Ward SR, Havlioglu N, Thomopoulos S, Galatz LM. Chronic degeneration leads to poor healing of repaired massive rotator cuff tears in rats. The American Journal of Sports Medicine. 2015;**43**(10):2401-2410

[53] Arvinius C. Estudio de la reparación de las lesiones del manguito rotador mediante compuestos transportador-BMP. [Tesis Doctoral]. Universidad Complutense de Madrid; 2018

[54] Esteban PT, Hoyas JA, Villafuertes E, Bobada CR, Tortuero GVG, Morales YL, et al. Tratamiento comparado de las lesiones del manguito rotador mediante el uso de células madre mesenquimales en combinación con membranas de colágeno tipo I. Trauma. 2012;**23**(Suppl. 1):70-75

[55] McCormick F, Gupta A, Bruce B, Harris J, Abrams G, Wilson H, et al. Single-row, double-row, and transosseous equivalent techniques for isolated supraspinatus tendon tears with minimal atrophy: A retrospective comparative outcome and radiographic analysis at minimum 2-year followup. International Journal of Shoulder Surgery. 2014;**8**(1):15-20

[56] Derwin KA, Baker AR, Codsi MJ, Iannotti JP. Assessment of the canine model of rotator cuff injury and repair. Journal of Shoulder and Elbow Surgery. 2007;**16**(Suppl. 5):S140-S148

[57] Ciampi P, Scotti C, Nonis A, Vitali M, Di Serio C, Peretti GM, et al. The benefit of synthetic versus biological patch augmentation in the repair of posterosuperior massive rotator cuff tears: A 3-year follow-up study. The American Journal of Sports Medicine. 2014;**42**(5):1169-1175

Chapter 2

# Experimental Animal Models for Studying Intestinal Obstruction

*Eleftheria Mavrigiannaki and Ioannis Georgopoulos*


## Abstract

Understanding the pathophysiology of intestinal obstruction and exploring potential therapeutic interventions heavily relies on the utilization of experimental animal models. This book chapter provides a comprehensive overview of various animal models employed in the study of surgically induced intestinal obstruction. From rodents to large animals, a range of experimental setups and methodologies are discussed, each offering unique advantages and insights into the complexities of this condition. The chapter provides a guide for researchers aiming to investigate intestinal obstruction, reviewing all aspects of an experimental protocol. Ethical and regulatory regulations, anatomical and physiological differences among species, and surgically induced experimental intestinal obstruction are reviewed. The existing experimental animal models are evaluated regarding their reproducibility and efficacy, as many published models prove difficult to replicate in the laboratory or lack crucial information for researchers. A chapter of this nature would greatly benefit the research community and pave the way for future studies.


**Keywords:** intestinal obstruction, animal models, experimental intestinal obstruction, experimental surgery, types of intestinal obstruction

## 1. Introduction

Medicine has acknowledged intestinal obstruction (IO) as a clinical entity for thousands of years and has used experimental research ever since to bring light to our understanding regarding the symptoms, pathophysiology, and management. Ebers Papyrous was the first to present a definition in 1550 BC, and Hippocrates was the first to state a treatment for ileus {ειλεός/eileós}, meaning the squeezed or twisted bowel [1, 2]. Nowadays, IO is characterized by the level, the degree, and the possible cause of obstruction. Thus, it might be of intraluminal, extraluminal, or intramural cause, either partial or complete, and located anywhere across the gastrointestinal tract (GIT). The inability of the intestine to perform normal propulsion of its content, either liquid, solid, or gaseous, triggers adjusting mechanisms to overcome the site of obstruction, such as proximal distension of the intestinal lumen and distal collapse, inhibition of normal secretory and absorptive function and cease of normal gastrointestinal (GI) reflexes. Complete intestinal obstruction is a surgical emergency, whereas conservative management may be therapeutic for partial intestinal obstruction, particularly when located in the small bowel, with success rates reaching up to 90% [3].

Prompt treatment is crucial for prognosis in all types of IO as they pose a major cause of abdominal morbidity and mortality [3–6]. The overall burden of IO has risen significantly in the last 30 years, from 56.91% in 1990 to 86.67% in 2019. There are geographic disparities of the age, type, sex, and pattern of IO due to the wide spectrum of pathologies hindered behind this clinical entity and the different socioeconomic, cultural, and dietary profiles, yet the overall rising trend establishes a significant public health issue [7]. Thus, research on IO is an ongoing field of investigation.

Animal models have already had a major contribution to our understanding, regarding the clinical entity of IO. Animal models offer the unique ability of *in vivo* real-time study of complex physiological and pathological phenomena that are both challenging and ethically impossible to replicate in humans. In vivo research offers the ability to study the sequela of any IO on homeostasis and the function of other organs as well as evaluate the efficacy and safety of surgical and non-surgical treatments and drugs. Experimental animal models are also important mediators in the training of physicians to acquire and advance surgical skills and diagnostic modalities. On the other hand, the advent of medical and research ethics has led to a series of legal acts that protect animals' rights and welfare and restrict the use of animal testing to cases involving the yet undefined pathophysiological aspects of intestinal obstructions and where viable and translatable alternatives to *in vivo* animal experimentation are unavailable.

In published literature, there is a large volume of experiments using one or multiple techniques of IO. All of these experiments are very heterogeneous as study protocols examine different aspects of IO and related preventive, therapeutic, or diagnostic insights. Unfortunately, most techniques are described briefly and are not accompanied by images of the procedure or comments regarding intraoperative difficulties or adverse events. For young researchers, all of these aspects might lead them to tough spots in the field. The main drawback, though, is the lack of comparative studies regarding the reliability and reproducibility of experimental intestinal obstruction techniques published by individual researchers.

Even though many aspects of the pathophysiology of IO have been clarified by previous experiments, there are still aspects of IO in healthy subjects or in combination with intestinal pathologies that require further research and perhaps experimental studies. The authors of this chapter encountered all the difficulties of designing a research protocol for experimental IO. Georgopoulos et al. [8] based on clinical observation, aimed to test the never-before-proven hypothesis that the expression or recurrence of Crohn's disease in humans is related to narrow intestinal passings such as valves, strictures, or tight anastomoses. To prove this hypothesis, the authors set up an experimental research protocol using knock-in mice that combined partial intestinal obstruction and genetic predisposition in Crohn-like disease. Yet, none of the existing methods of partial IO could fulfill the criteria of a long-standing partial IO. The authors decided that a research protocol to define the most suitable model had to be preceded. After reviewing and replicating the existing methods of partial IO, Georgopoulos et al. [9] established a new, robust, reproducible, and refined model of partial IO, ultimately pioneering a novel triple suture technique. This research gave very interesting insights into Crohn's disease, which has still many obscure pathophysiological pathways, showing evidence that this clinical observation may actually be accurate.

In the following chapter, we aim to provide researchers with a guide for proper protocol planning and decision-making regarding an experimental surgical intestinal obstruction study from scratch.

## 2. Choosing the most appropriate animal model

One of the parts that need thorough and in-depth study at the time of shaping a research protocol is the part of deciding the appropriate animal model for the translational research project at hand [10]. The aspects that need to be taken into consideration are summarized in **Table 1**.

### 2.1 Ethical and regulatory considerations

Publications in esteemed journals, as early as the nineteenth century, show interest in the rights of experimental animals [11, 12]. Russell and Burch were the first to use the terms "replacement, reduction, and refinement" in 1959 [13], and a little later, the "3Rs principle" used by Smyth [14]. The 3Rs comprise a series of measures to be taken in order to perform animal experiments in medicine in a more humane way [15, 16].

The principle of the 3Rs is explained briefly in **Table 2**.

These principles have been widely accepted and are used as a guide for experimental research worldwide, even though they are not implemented to the same degree in all countries and by all researchers [17]. Initiated by the National Centre for the Replacement, Refinement, and Reduction of Animals in Research (NC3Rs), a checklist of 20 items, the ARRIVE guidelines (Animal Research: Reporting of In Vivo experiments), was suggested in 2010 in order to help authors and journals identify and include the minimum information necessary to report in publications describing in vivo experiments [18]. This checklist includes information concerning the characteristics of

| |
|---|
| 1. Ethical and regulatory considerations |
| 2. Availability of animals and housing facilities |
| 3. Translational research suitability of animal species |
| 4. Surgical and anesthetic considerations |
| 5. Overall costs of the study |

**Table 1.**
*Crucial aspects to consider when choosing the appropriate animal model.*

| | Basic | Updated |
|---|---|---|
| Replacement | Avoiding or replacing the use of animals in areas where they otherwise would have been used. | Accelerating the development and use of predictive and robust models and tools, based on the latest science and technologies, to replace the use of animals in addressing important research questions. |
| Reduction | Minimizing the number of animals used is consistent with scientific aims. | Appropriately designed and analyzed animal experiments that are robust and reproducible and add to the knowledge base. |
| Refinement | Minimizing the pain, suffering, distress, or lasting harm that research animals might experience. | Advancing laboratory animal welfare by exploiting the latest *in vivo* technologies to minimize pain, suffering, and distress and improving understanding of the impact of welfare on scientific outcomes. |

**Table 2.**
*Definitions of 3Rs – basic and updated (by the National Centre for the Replacement, refinement, and reduction of animals in research – NC3Rs [13]).*

animals used, details on housing and husbandry, details on drugs or surgical procedures applied and statistical and analytical methods used, including methods used to reduce subjective bias by the researchers [18]. The ultimate aim was to tackle the problem of serious omissions during experimental research and their potential scientific, ethical, and economic implications. Due to the limited effect on the transparency of reporting in animal research since their first publication in 2020, these guidelines were updated to further facilitate their use in practice [19]. Many national and international regulations and directives have encompassed these guidelines [20, 21], but still compliance with the 3Rs principle is not achieved to the desired extent [22].

### 2.2 Availability and housing facilities

Animals for research may be either purchased from licensed suppliers or provided from the in-house bred colonies of the research center. Depending on the content and the power analysis of the study, a certain number of animals of certain characteristics (age, sex, weight, etc.) will be needed throughout the study. The researcher should design a detailed timeline for the study, explaining the number of animals needed at certain times, the duration of their housing, the timing of each intervention (surgical, veterinarian, genotyping, breeding, etc.), and make all necessary arrangements to have the available animals throughout the study at the times needed [12, 23]. The researcher should beforehand make sure of the appropriate housing conditions accordingly to the animal species chosen and find the appropriate facility to house the study. Housing conditions such as cages, boxes, or even shelters and outdoor housing in larger animals, are important for the well-being of the animals. There are recommendations for the minimum space in the animal house for every animal species, so care should be shown to the availability of appropriate space for the number of animals being housed at the same time during the study. Special attention should be given to assure healthy environmental conditions, such as the circle of light and darkness, the noise, the temperature, and the humidity, and that adequate and proper – for the study and species – husbandry is readily available [12, 15].

### 2.3 Translational research suitability of animal species

Translational research is the process that leads from findings of basic science (ideas, insights, discoveries, etc.) to bedside medicine, such as novel treatments, prevention measures, and a better understanding of human disease. A basic step in deciding the appropriate animal model for a research protocol is choosing among the different animal species available for the one that has a higher potential translational impact. In other words, looking for an animal that has a greater genetic, anatomic, physiologic, pharmacologic, biochemical, etc., resemblance or comparability to human.

Both large and small animals have been used as subjects to study IO. Rodents and pigs have prevailed over dogs, monkeys, sheep, rabbits, and horses in experimental models of IO. Hatton et al. [24] reported that from 1966 to 2014, rodents consisted up to 90% of animal species in biomedical research, with rabbits, pigs, dogs, monkeys, and guinea pigs accounting for the rest 10%. The same trend toward rodents is underlined by Oliveira et al. [25], reporting that 73% of ischemia-reperfusion models utilize rodents as animal species, followed by pigs at 9%, dogs at 8%, and felines at 6%. Ewes' models are mostly cited in the literature in the mid-twentieth century, an era when metabolic mechanisms hindered behind electrolyte changes and clinical outcomes were under the microscope, as well as research on the field of bowel response to trauma and pain [26–28]. Only 15% of sheep used in experimental research have served as models for gastrointestinal studies as they are herbivores with significant anatomic and physiologic differences compared to humans [10]. Non-human primates, dogs, cats, and Equidae have been used due to the better resemblance of their anatomy and physiology to humans, yet due to ethical concerns, they are protected by the "Animals Scientific Procedures Act," and their use in experimental medicine is allowed only when other species are not eligible as animal models for a certain aim of study [29, 30]. Parameters guiding the proper animal species, in general, are dictated by the aim of the study and the ethical, legal, and regulatory requirements that protect animal welfare, which are not globally equal and are customized by each country's constitution.

### *2.3.1 Small animals*

Small animals such as rodents and rabbits, are readily available animals, easy to handle, and more cost-effective in terms of procurement, housing, and maintenance compared to larger animals (**Figure 1**). Laboratories with limited facilities have the space requirements to host small animals in large samples. Rodents are nocturnal animals which needs to be interpreted in their housing conditions. Rabbits newly introduced to a colony should undergo a mandatory quarantine period that allows the rabbits to adapt to their new environment. Providing rabbits between 12 and 14 hours of light ensures synchronization with the colony's circadian biorhythm [31]. Genetic manipulation techniques are well established in small animals; thus, researchers have a wide range of modified animals to study specific genetic factors related to intestinal obstruction and explore potential therapeutic targets. Their short reproductive cycles allow for the generation of a large number of experimental subjects in a relatively short time. Gestation lasts approximately 19–21 days for mice with a litter size of 7–11 young, 21–23 days for rats with a litter size of 6–13 young, and approximately 30–32 days for rabbits with a litter size of 5 to 8 young [32]. This is particularly beneficial for studies requiring a significant sample size. On the other hand, the small size may restrict the feasibility of certain surgical procedures and utilizing imaging techniques, making it challenging to replicate complex clinical scenarios related to intestinal obstruction. Adult mice have an average weight of 20–40gr for males and 22–63gr for females. Rats are larger animals than mice, with an average adult weight of 267–500gr for males and 225–325gr for females, while rabbits weigh up to 3 kg at

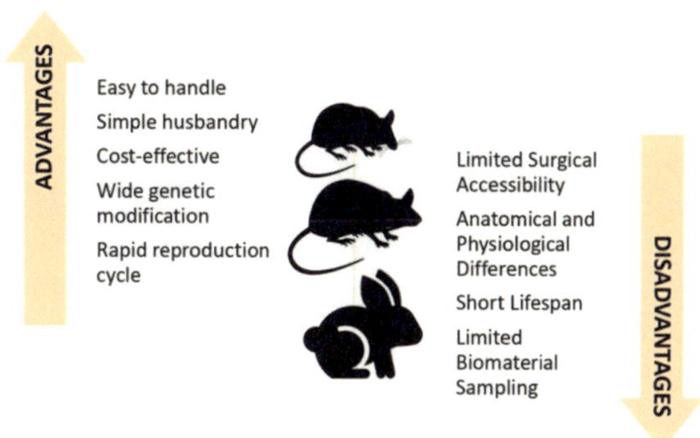


**Figure 1.**
*Main advantages and disadvantages of small animal models in experimental IO.*

5 months of age [31]; thus, surgical manipulation is more feasible in rats and rabbits than in mice. Small animals generally have shorter lifespans, limiting the duration of long-term studies and the ability to observe the chronic effects of intestinal obstruction. The average life span of mice is 12–36 months, and of rats is 26–40 months [32]. Gut-associated lymphoid tissue is distributed homologously in rodents and humans by means of quality and quantity [24]. Rodents' genome is approximately 14% smaller than humans. These models offer several advantages, including ease of genetic manipulation, particularly in immunodeficient mouse models, cost-effectiveness, and the availability of well-established genetic tools. However, despite these benefits, it is important to acknowledge that there are significant differences between humans and rodents. These differences pose limitations in accurately simulating complex diseases and translating research findings to clinical practice.

### *2.3.2 Large animals*

Large animal models in experimental intestinal obstruction allow for a more realistic simulation of surgical procedures and render the use of surgical instruments more feasible. It is also more feasible to utilize advanced imaging techniques that provide valuable insights into the structural and functional changes associated with intestinal obstruction. Experimental scales are, in general, more comparable to humans in terms of intervention, clinical scenarios, and treatments; thus, they are also the preferred models in hands-on teaching seminars. Their longer lifespan enables researchers to extend observation periods regarding the progression of IO and its outcomes and the conduction of longitudinal studies. Pigs have a lifespan of about 27 years, and dogs of 10–13 years. Gestation in pigs lasts 114 days with a litter size of 12 piglets [33]. Porcine adulthood is reached by the first year of life. The first 6 months of life represent childhood and puberty, which lasts from 6 to 12 months of age [34]. Adulthood in dogs is reached approximately the second year of their life, and they are considered adolescents from 6 months to 24 months of age. Childhood is considered the period from birth till the sixth month of age. An important aspect is that large animal species' gastrointestinal anatomy and physiology are widely similar to humans; thus, they are more representative models for investigating all types of effects of IO. The porcine genome is proven to be three times more homogenous to humans than the mouse genome, and the porcine immune function approximates humans by 80%, although there are variations in the porcine adaptive immune response [35]. The primary drawback of utilizing large animal models, besides ethical considerations, is the elevated expenditure associated with their maintenance and care. Another important limitation is the limited availability of genetically modified models. Larger animal species necessitate more extensive and specialized housing and surgical facilities, resulting in increased costs for feed, veterinary care, and surgical procedures (**Figure 2**). There is a scarcity of studies exploring the manipulation of the dog genome [36]. Genetic manipulation in pigs is not as developed as in rodents but has produced an ideal analog to human cystic fibrosis. Besides domestic pigs, there are 29 different lines of pigs, wild type and genetically engineered, available as experimental models [35].

### *2.3.3 Interspecies anatomy and physiology*

Anatomy and physiology are highly correlated to species evolution and vary depending on sex, age, diet, weight, and genome, not only among animal species

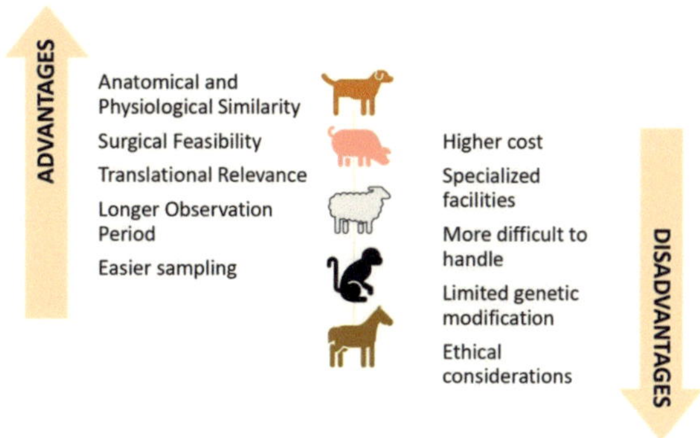


**Figure 2.**
*Main advantages and disadvantages of large animal models in experimental IO.*

but even between breeds. All these parameters are crucial when designing a study protocol and even more so when matching results to human conditions. Kararli et al. [37], Hatton et al. [24], and Zwart et al. [38] have published in-depth descriptions of the comparative anatomy and physiology of the human intestine and commonly used animal species.

Pigs and dogs are omnivore monogastric species with glandular-type stomachs like humans. Rodents are omnivorous monogastric species with both glandular and non-glandular stomachs separated by a bridge, the margot plicatus [38]. Their inability to vomit is attributed to this bridge in combination with their high metabolic rate; thus, no fasting is needed before anesthesia. Rabbits are herbivorous monogastric species with simple stomachs that lack specialized regions and are thin-walled (**Figure 3**). Rabbits, as rats, are also cecotrophic, meaning that they re-ingest fecal material. Rabbits, as rodents, are continuous feeders with the inability to vomit; thus, no fasting is required prior to anesthesia. Gastric transit time is 3-6 h, cecal material remains in the fundus of rabbits for 6-8 h, and normally, the stomach contains a mixture of food, hair, and fluid even after 24 hours of fasting [39]. Gastric emptying is incomplete in pigs, so they tend to retain food in the stomach even 24 hours later [38]. Cyclic gastric motility has the same phasic pattern and duration per phase in dogs as in humans and lasts approximately 2 hours [38].

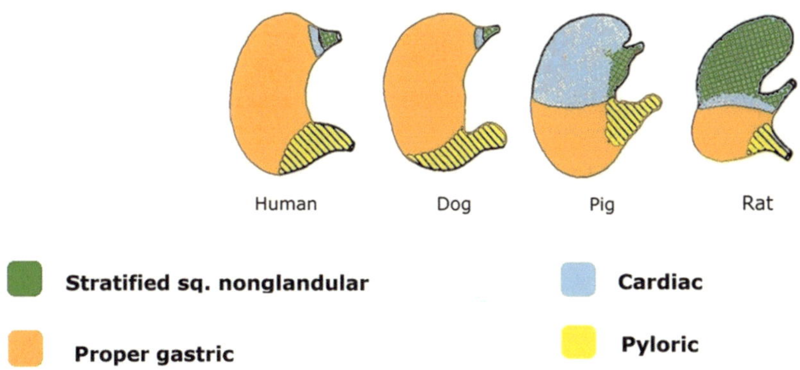


**Figure 3.**
*Interspecies gastric mucosal variations. (interpreted image by Kararli et al. [37]).*

Humans and rabbits typically exhibit few organisms in the upper GI tract due to lower pH values in the fasting state, while other animals often harbor large numbers of bacteria in the stomach and upper intestine (**Table 3**). pH value variations along the GIT tract affect drug absorption and metabolism and need to be interpreted when translating research outcomes to human conditions (**Table 4**). Microflora exhibits lower heterogeneity in the lower intestine in all animals, including humans. While humans harbor a predominantly consistent population of Bacteroidetes and Firmicutes phyla in the GI tract, animals exhibit diverse gut microbiome compositions based on their dietary habits 85% of mouse gut microbiota are absent from human flora [40]. These microbiota variations play critical roles in nutrient absorption, fermentation processes, and disease modulation. Additionally, gut microbiota influence drug metabolism and absorption, with regional concentrations varying across species. Nonetheless, modified animal models, like germ-free rats, offer insights into drug stability and microbial interactions. Mouse models, sharing similarities with human microbiomes, are extensively utilized for host genetics and microbiome studies, facilitating extrapolation of data.

The measurements of intestinal length in these studies refer to cadaveric observations, and as noted by the authors, measurements are longer than in vivo length. The jejunum is the longer part of the GIT in most animal species. **Figure 4** shows the gross GIT anatomy of the most commonly used animal species, and **Table 5** summarizes the differences in intestinal length between the human intestine and the most commonly used animal species. Porcine intestinal length is 20-fold longer than that of humans, approximately 23 m (range 15-24 m) [37], but if the size of intestinal length in meters is adjusted to body weight in kilograms, it results in a ratio of approximately 0.1 which correlates to the human ratio. The small intestine

| | Human | Mouse | Rat | Rabbit |
|---|---|---|---|---|
| Stomach | 0–5 | 7–9 | 7–9 | 0–6 |
| Proximal small intestine | 0–5 | 7–9 | 6–8 | 0–5 |
| Distal small intestine | 6–7 | 7–8 | 7–8 | 6–7 |
| Colon | 7–10 | 8–9 | 8–9 | 8–9 |
| Rectum/feces | 10–11 | 9–10 | 9–10 | 9–10 |

**Table 3.**
*Distribution of gut flora across different species in bacterial counts at different segments of the gastrointestinal tract quantified as the logarithm of viable organisms per gram of wet weight. Interpreted by Kararli et al. [37].*

| | Human | Pig | Dog | Mouse | Rat | Rabbit |
|---|---|---|---|---|---|---|
| Stomach Fasted state | 0.4–4 | | 1.5 ± 0.04 | 4.04 | 3.9 | |
| Stomach fed state | 1.5 | 4.5 | 3–5 | 2.98 | 3.2 | 1.5 |
| Jejunum | 6.6 ± 0.5 | 6.2 | 6.2 | 5.01 | 6.13 | 6.8 |
| Ileum | 7.5 ± 0.5 | 6.9–7.5 | 6.6–7.5 | 4.8–5.2 | 5.93 | 7.5–8 |
| Colon | 6.4 ± 0.6 | 6.8 | 6.5 | 5.02 | 6.23 | 7.2 |

**Table 4.**
*Interspecies pH values of the GIT incorporated by Hatton et al. [20].*

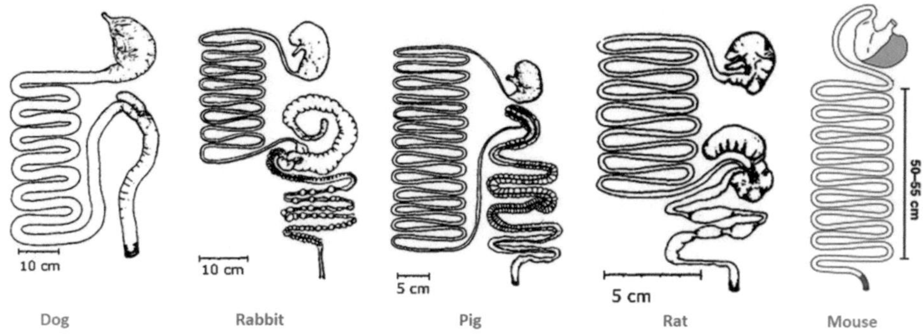


**Figure 4.**
*Differences in the macroscopic GI anatomy among species (incorporated image by Kararli et al. [37] and Ziegler et al. [40]).*

| | Human | Pig | Dog | Mouse | Rat | Rabbit |
|---|---|---|---|---|---|---|
| Average total intestine length (m) | 7 | 23.5 | 4.82 | 0.5–0.55 | 1.2–1.7 | 5.82 |
| Small intestine (m) | 6.25 | 14.16 | 2.48 | 0.34 | 1 | 3.39 |
| Cecum | 0.15 | 0.23 | 0.08 | | 0.04 | 0.44 |
| Appendix | + | — | — | — | — | + |
| Colon(m) | 1.5 | 4.27 | 0.6 | 0.09–0.142 | 0.26 | 1.23 |

**Table 5.**
*Interspecies intestinal length interpreted by Hatton et al. [24], Kararli et al. [37], Zwart et al. [38], Ziegler et al. [40].*

is twice the size of the human, approximately 14.5 m long, and the colon is 4.7 m. The caliber of the small intestine is around 3.5 cm, which is close to the human caliber and shares a similar finger-shaped architecture of the microvilli. The main differences between porcine models are the lack of an appendix, the enlarged cecum, the spiral configuration of the large bowel, and the inverted lymph node structure with lymphoid follicles present as meter-length strips [24, 35, 41]. The canine anatomy and physiology are more similar to humans than any other species. Canine intestinal length is half the size of the human intestine, approximately 4 m in total length [37]. Canine body length of 0.75 m correlates to 3.9 m of small intestine 3.9 m and 0.6 m of large intestine. The small intestine has longer and more slender villi, creating a capacious absorptive surface area with an estimated 34 microvilli per micrometer of villi and 23 villi per square millimeter of freeze-dried dog jejunum and ileum [24]. The diameter of the small intestine varies between different canine breeds. The cecum is proportionally equal to human cecum but coiled and appears as a diverticulum to the right of the ascending colon [24]. The main anatomical differences of canine models are the absence of an appendix, the absence of haustra on the large bowel, the shorter mesentery, which results in a relatively fixed position of the colon, and the absence of sigmoid colon.

The total intestinal length of mice is approximately 0.5-055 m. The small intestine is approximately 0.34 m long and has a finger-shaped architecture of the microvilli, as seen in pigs and humans. The diameter of the small intestine ranges from 2.3 mm to 2.9 mm. The cecum is the most enlarged part, with a diameter of 4.9–5.4 mm, and has no appendix [42]. Their colon is approximately 0.09–0.14 m long with a diameter of 2.7–2.9 mm, lacks haustra, and it is not clearly divided into sections as the human colon [43].

Rats' intestinal length is approximately 1.2–1.7 m. The small intestine is approximately 1–1.4 m with a diameter of 3–5 mm. The rat colon diameter is approximately 10 mm [37]. Microvilli has a tongue-shaped architecture with 65 microvilli per square micrometer of villi surface. Rats' colon is approximately 0.22–0.26 m, lacks haustra, and its first part is structured like a cecum. The sigmoid colon is absent, and so is the gallbladder. Rabbits' intestinal length is 5.8 m. The small intestine is approximately 2.70 m, and the large intestine is 1.95 m with a diameter of 0.3–0.5 mm. The jejunum is the dominant region of the small bowel and ends up with a well-defined appendix rich in lymphoid tissue [24]. The cecum in the rabbit is a well-defined, thin-walled, and coiled structure that can hold up to 40% of the GIT content, meaning it is 10 times more capacious than their stomach.

## 2.4 Surgical and anesthetic considerations

### *2.4.1 Surgical considerations*

A series of surgical and anesthetic factors should be assessed by the research team (e.g., the surgeon, veterinarian, animal care staff, and the investigators) before concluding with the preferred experimental model. The presence of a multidisciplinary team in a surgical project seems to increase the likelihood of a successful outcome [44].

The surgical procedures are categorized as major or minor depending on the physical or physiological impairment or the presence of extensive tissue removal or other major alterations. Surgical procedures involving the intestine may be major if the effect is for example total intestinal obstruction. Animals recovering from major surgical procedures may require extra care for pain, discomfort, and complications, close postoperative monitoring to assure animal well-being, or euthanasia if pain or distress cannot be alleviated medically. Minor surgical procedures usually have minimal complications, and animals shortly return to normal function [15, 45].

The researcher should try to choose among surgical models that are reliable and reproducible. Many literature articles, as stated elsewhere in this chapter, fail to report all the needed information in order to be reproduced, and many seem never to have been replicated by other researchers. Reliable models are those that show results in correlation to the certain surgical effect inflicted. Reproducible models are those that have been replicated by other researchers with the same or comparative results [23].

Nevertheless, all surgeons to be involved in a study should be of adequate expertise for the certain procedure and agree to every detail relevant. The complexity and difficulty of the surgical procedure or intervention chosen should be relevant to the dexterity and expertise of the team. Surgical training in non-living models such as artificial simulators, animal tissues, animals euthanized for other reasons, and pilot studies may be required before commencing the experiment, in order to improve the available skills on a certain animal species, test the feasibility of the procedure and assure the consistency of the team [15, 23].

Surgical outcomes should be monitored continually and assessed by the research team for the protection of the well-being of the animals and in order to make appropriate changes in time for a successful outcome [15]. Other surgical details such as the use of aseptic technique, the preparation of the surgeon, and the sterility of instruments and materials should be answered, since they may create extra logistic concerns. The aseptic technique is widely accepted to reduce the likelihood of infection, though some small species such as the mouse or rat, show no complications with just clean surgical conditions [9, 15, 46].

### *2.4.2 Anesthetic considerations*

Detailed guidelines and current best practices in the use of anesthetics and analgesics are outside the scope and purpose of this chapter. The researcher should seek this species-specific information in laboratory guides or veterinary directives according to the design, the available equipment, and team expertise. We are limited to a few general principles and tips that are potentially helpful for the researcher.

#### *2.4.2.1 Analgesia*

It is generally accepted that animals experience pain in a similar way to human. Pain in animals may affect the experimental results by delaying recovery, anorexia, altering their mental status, creating major endocrine responses, and more. Thus, scientific reasons, as well as ethical reasons, compel the use of analgesics in surgical research [12, 47, 48]. The regimens, the dosage, and the indications or contradictions depend on the animal and the surgical intervention and should be individualized in every experiment [12, 47–49]. Preemptive analgesia (preoperative and intraoperative analgesia) seems to be beneficial in optimizing the intraoperative and postoperative course of the animal and perhaps should be implied in experimental protocols with major surgical procedures [47, 49–51]. Post-operative analgesia should always be administered until at least the restoration of food and water intake and resuming of the normal behavior of the animal. Frequent monitoring of the animals' behavior port-operatively will indicate the timing and length of treatment [50]. Even though simple analgesics usually do not interfere with the scientific findings and are widely used in research, special care should be given with regard to the effects of certain agents on certain tissues to avoid any unwanted confounding factors.

#### *2.4.2.2 Anesthesia*

The type of anesthetic agents used in an experimental study is usually guided by numerous factors such as the expected length of the surgical procedure, the available equipment of the operating facility, the expertise of the team, the presence of veterinarian, the animal species and first and foremost the accepted current practices in regard to animal welfare [47, 50]. The two main anesthetic regimen categories are injectable and inhalant. Injectable anesthetics can be administered through intramuscular, intraperitoneal, or intravenous injection. Inhalant anesthetics can be used by anesthetic chamber or face mask and maintained using a face mask (in smaller animals especially) or an endotracheal tube. Although inhaled anesthesia is considered safer than injectable, its use in literature is limited to larger animals due to lack of equipment or expertise of the researchers [47, 50, 52].

Intraoperative monitoring of anesthetic depth and vital signs, such as body temperature, cardiac rate, blood pressure, and respiratory rate, increases the likelihood of a successful surgical outcome. The recovering period post general anesthesia is of high risk for advent events that may be lethal. Continuous attention should be given until the effects of the anesthetic regimens have worn off, and the researchers should be vigilant for any supportive treatment to be given timely. During that time, it is perhaps the best time to administer the first dose of post-operative analgesia to benefit from the long-term effects of analgesia and less adverse events during recovery [50].

*2.4.2.3 Euthanasia*

The researchers should include in their study details of the euthanasia procedure, considering that practically all experimental animals will eventually be killed before their life span, with very few exceptions. The timing of killing may be scheduled beforehand by the design of the study, such as collection of tissues or reaching the end of their breeding life or stock use. Animals, though, should be killed if they are in severe pain or suffering without possibility of recovery (Article 8 of Council Directive no. 86/609/EEC). Killing an animal should be always carried out in a humane way [12, 20]. There are several ethically and scientifically acceptable methods of euthanasia available that are individualized for each species [47]. Two widely acceptable ways for most animal species are overdose of injectable anesthetic agents and overdose of anesthetic agents by inhalation with exposure to carbon dioxide [45, 48, 51, 53].

Before the study begins, humane endpoints should be defined. These refer to the earliest indication that an animal, during an experiment, is in severe pain, suffering, or distress, or there are signs of impending death. The study team should describe humane endpoints in meaningful terms, establish observation schedules for prompt recognition and make all arrangements for provision of measures to alleviate these symptoms or signs at sight. In case of irreversible situations, the animals shall be euthanized without delay [45, 50].

### 2.5 Overall cost of the study

The overall costs of the study are usually a definitive factor in an experimental study. In order to calculate a realistic budget for each study, it is crucial to have all the available information and costs from every aspect of it, which can be done after in-depth study and detailed design. Some important costs to take into consideration -where applicable- are the cost of purchasing the total number of animals, the cost of maintenance, housing, and husbandry per day and animal, the cost of drugs and surgical tools or materials, the cost of tests on the animal (blood, stool, urine, genetic, etc.), the cost for disposing of the animals according to international regulations and the cost of any special studies on the animals or the specimens (molecular, histologic, nuclear, radiologic, etc.) [23].

## 3. Types of experimental intestinal obstruction

All surgical procedures (**Figure 5**) must be preceded by proper housing, fasting, anesthesia, and prepping of the animals. Basic surgical skills are required in order to safely enter the abdominal cavity and handle the intra-abdominal tissues so as to prevent surgical complications and reduce perioperative mortality. Basic knowledge of instruments and tissue handling is also essential for researchers who aim to investigate an experimental IO.

### 3.1 Mechanical intestinal obstruction

Mechanical IO refers to an anatomic barrier that might be either partial or complete. In complete IO, the propulsion of intestinal contents is totally halted beyond the site of obstruction, whereas in partial IO, some gaseous or liquid enteric content is able to pass distally. Mechanical IO is further divided into acute or chronic, depending

| EXPERIMENTAL SURGICAL INTESTINAL OBSTRUCTION MODELS | | |
|---|---|---|
| MECHANICAL INTESTINAL OBSTRUCTION | | FUNCTIONAL INTESTINAL OBSTRUCTION |
| COMPLETE | PARTIAL | |
| Suture Ligation | Intestinal ring | Intestinal manipulation |
| Clip obstruction | Simple ligation | Intraperitoneal irritation |
| Clamp obstruction | Tripple suture ligation | Cecal ligation and puncture |
| Balloon obstruction | Intestinal clip | Two – hit technique |
| Closed loop ligation | Tube Ligation | Ischemia - Reperfusion |
| Adhesion models | Adhesion models | |
| Mesenteric ischemia | Intestinal valve | |
| | Stricture | |

**Figure 5.**
*Experimental surgical intestinal obstruction models.*

on the time of onset of symptoms. The underlying pathology might be extrinsic such as adhesions, hernias, intra-abdominal masses, and volvulus, and intrinsic such as intestinal neoplasmatic lesions, inflammatory diseases, congenital malformations, duplications, or intraluminal, including intussusception, foreign bodies, hematomas, gallstones, and feces [3, 54, 55].

### *3.1.1 Complete mechanical obstruction*

*Ligation models:* A simple ligation of the lumen at the desired level of obstruction has been a commonly cited technique since the beginning of the twentieth century [26–28, 56–73] and is probably the most feasible and cost-effective model to produce an acute complete intestinal obstruction. The desired intestinal loop is identified after entering the abdominal cavity, and the adjacent mesentery is carefully dissected so as to create a small mesenteric window but not to damage the mesentery itself or the mesenteric marginal vessels. A suture ligation is passed around the intestinal lumen and tied. Silk sutures are most commonly used, but any type of suture material is acceptable [59]. This technique is easy to perform even from researchers with basic surgical experience. The main drawback is that the pressure force of the ligating suture cannot be controlled, and neither is reported using measurable values. An overly tied ligation will crush the intestinal lumen and result in rapid necrosis. Enochsson et al. [74] reported a loose ligation around the intestinal lumen of rats, yet this is not an objectively measurable value. Many researchers note that they place the ligature in such a way so as not to strangulate the intestinal lumen. This is not yet proved by objectively measuring different levels of tightness.

Other methods used to generate an acute complete obstruction utilize clips, automatic mechanical suturing devices, rings of various materials such as silicone tourniquet catheters, GoreTex bands, cotton threads, umbilical tapes, and clamps placed around the intestinal lumen or in Ω shape [46, 75–87]. Transection of the bowel, either via open or minimally invasive approach, is also reported in order to induce an acute complete IO [88]. Correa-Martin et al. [85] utilized a forced laparoscopic suture at the ileum near the ileocecal valve of 10 large female pigs, reporting only one death due to bowel perforation. Berlin et al. [46] used a ligating clip to induce small bowel obstruction in 19 male rats, noting that tight apposition was avoided to prevent local necrosis without recording intraoperative adverse events. Zhang et al. [80] created an Ω shape clamp from a double flat iron-core binding wire of 3 mm width in order to produce a more graduate, stable, and reversible complete obstruction of the small intestine in rats. A total of 200 rats were used of which 60 were assigned to recovery groups after de-obstruction, reporting 10 deaths and minimal intestinal injury by the clamps. An obstructive device was developed by Fraser et al. [89, 90] by attaching a silicon rubber sheet at the balloon tip of a silicone rubber urethral catheter. The dimensions of the materials used to produce the device were adjusted to the predicted intestinal lumen of the three species they used: 5 monkeys, 7 dogs, and 20 rats. The device was wrapped around the intestinal lumen, the edges of the silicon sheet were sutured, and the free end of the catheter was fixated to the abdominal wall. As reported by the authors, they were able to predictably and progressively replicate both acute and chronic complete IO by changing the insufflation level of the balloon.

*Strangulation models* [78, 91–99]: By definition, strangulation refers to a disruption in the blood supply to the intestine. Vascular occlusion can be achieved either by ligating or clamping the vessels. A simple technique to strangulate the intestine is to ligate an intestinal loop with its adjacent mesentery by passing a simple ligating suture around its basis so as to create a "fan" from the loop and the mesentery. Another common technique is ligating separately an intestinal segment and its mesenteric feeding vessels or the main feeding vascular branch. This is called the segmental mesenteric vascular occlusion by Gonzalez et al. [100] while other authors report this as the closed-loop strangulation technique [101–103]. This model can be performed both in large animals and rodents, providing multiple treatment groups in a single animal as multiple loops can undergo varying durations of ischemia, with or without reperfusion, allowing for adjustment of the degree of injury. Oliveira et al. [25] in order to precisely locate the marginal artery branch, injected blue liquid dye in the jugular vein and carotid artery of a "pilot" rabbit before the experiment and managed to establish consistency of the occlusion site. Matsuo et al. [104] placed an acrylic ring around a balloon and a loop of the distal ileum of rats. By inflating the balloon, they managed to obstruct blood flow to the adjacent intra-ring loop, followed by deflation to reverse the strangulation. Fevang et al. [105] placed an infant blood pressure gasket around a loop of porcine ileum and increased the pressure until venous pressure reached 50 mmHg and managed to simulate strangulation. This technique is not applicable to small animals due to size restrictions. Manual rotation has been utilized by Doğan et al. [106] and Darien et al. [107] in order to simulate the volvulus of the small intestine in rats and the ascending colon in ponies, respectively. All of these models can be used to study ischemia and reperfusion intestinal injury by reversing the occlusion of the blood supply strangulation models are valuable to study clinical scenarios such as volvulus or incarcerated hernias.

*Atresia models:* Ligation of mesenteric vessels that perfuse a part of the small intestine of animal fetuses was the first described technique to induce congenital intestinal atresia based on the theory of an intrauterine vascular insult being the pathogenetic mechanism [108–110]. Baglaj et al. [111] used a microsurgical bipolar coagulator instead of ligation to produce the vascular ischemic event in chick embryos. Researches also utilize the intestinal loop ligation model, using fine sutures 9–0, 10–0, or 11–0 to ligate an intestinal segment, to surgically induce congenital IO in rat, rabbit, chick, and sheep fetuses [110, 112–116]. These techniques require advanced surgical skills, advanced equipment, and specialized laboratories to offer optimum care to mothers and offsprings. Certain pharmacological agents or toxins such as adriamycin [117] when administered to pregnant animals during critical stages of fetal development, disrupt normal intestinal development and induce atresia-like abnormalities in the offspring. Genetic engineering techniques have also been used to introduce specific mutations or alterations in key genes associated with intestinal development and homeostasis. This approach allows researchers to create animal models with genetic predispositions to intestinal atresia and study the underlying mechanisms involved [118].

### *3.1.2 Partial mechanical obstruction*

Partial IO models exhibit great variation in literature regarding terminology. Partial or incomplete obstruction refers to difficulty passing intestinal content beyond a specific point, which could result from external pressure (such as from a tumor, aneurysm, and adhesion) or stenosis. Models to induce external pressure in order to induce partial IO are mainly modifications of the aforementioned techniques using rings of different materials, clips, and penetrating ligations placed on only a proportion of the enteric lumen [119–130]. Stenosis, on the other hand, specifically refers to a scenario where the intestinal lumen is reduced in diameter due to a lesion within the intestinal wall, characterized by localized narrowing, irregular muscularis, and thickened submucosa, which needs to be testified by histology. Partial IO might be the onset of a complete obstruction. Models of partial IO must justify a chronic obstruction and yield different degrees of incomplete obstruction or stenosis without crashing the intestinal lumen.

*Intestinal rings* [9, 76, 83, 123, 124, 126–128, 130–135]: The intestinal ring technique is commonly employed in research, yet there is great heterogeneity regarding the size of rings and recoding of the reported ratio of obstruction. The main aspects of the technique are cutting a strip of available material (silicone, polyethylene, polyurethane, plastic film, and penrose) to form a band and pass it around the desired part of the intestine through a mesenteric window, followed by edge-to-edge suturing in order to form a closed circumferential structure around the intestinal lumen. In vitro replication prior to in vivo experimentation is advised, as it might be forceful to penetrate the edges of the strip, depending on the material used, thus causing inadvertent complications. The main drawback of this technique though is the lack of a standardized ratio of ring to intestinal diameter to produce a consistent type of partial IO. Morel et al. [122] described the use of multiple adjustable rings to an isolated loop of jejunum, without recording of ring sizes and material and concluded that maintaining $\geq$60% of the original diameter allows for normal flow of the content while a diameter of $\leq$30% results in bowel rupture.

*Clips* [9, 136]: The clip technique is easy to perform and requires basic surgical skills in order to enter the abdominal cavity and locate the desired intestinal loop. No handling of the mesentery is necessary. Besides its feasibility, this technique is not regularly reported in partial IO models. As reported by Georgopoulos et al. [9] it is not a somatometrically adjustable technique and it is difficult to create the same degree of stenosis and apply the exact closure force needed in order to create stenosis but to not crush the tissue. Another interesting note is the rejection of the clip at the time of autopsy, which might result in rise in the number of animals used and experimental time.

*Ligation models:* Ligation models can be divided into "tube ligation" and "simple ligation." In tube ligation models, a tube is placed on the longitudinal axis of the desired portion of the intestine, and a ligation is passed around the tube and the intestine and tied. The tube is then removed, and the knot is left intact surrounding the intestine [9, 125, 137]. Different sizes and material of tubes have been reported in the literature. In a comparative study of partial IO by Yuan et al. [125], they concluded that the wide pipe group of their study was the most effective in replicating incomplete IO. This group used a pipe of 10 mm in length and 6 mm in width for "tube ligation" of rat ileum, yet animals were euthanized at 72 h post obstruction. Georgopoulos et al. compared experimental partial IO and reported mortality before 96 hours for all tube sizes, with only one animal surviving until the sixth postoperative day; thus, they concluded that it should be considered more of a complete obstruction model. "Simple ligation" [9] is penetrating the intestinal lumen on a proportion of its diameter. A non-absorbable suture is passed through the lumen, transmurally, on its antimesenteric border, and tied. This is also a technically not demanding technique as there is no dissection to the mesentery; it needs simple instruments and basic skills. Researchers must ensure obstruction of the enclosed tissue without crushing it as the knot is tied. Georgopoulos et al. [9] developed a triple suture technique, overcoming the vulnerability of the single partial ligation and achieving a long-term survival of animal models up to 4 weeks postoperatively. In the triple suture technique, three consecutive penetrating sutures were passed through the lumen, transmurally, on its antimesenteric border with a distance of approximately 1.5 mm between them.

Some authors have reported isolated partial IO models that are more challenging to replicate compared to the aforementioned techniques. Collins et al. [138] reported a technically challenging method on porcine models of creating a partial IO by combining a wide band of polytetrafluoroethylene placed around the intestine and folding the intestine telescopically on top of this band to create a valve mechanism. Besides being technically demanding, this technique also deems necessary at least 7 cm of intestine for its construction. Hiratsuka et al. [139] developed a porcine animal model of colonic stenosis using a silicon sheet (40 × 80 mm) onto which there were attached silicon rubbers at intervals of 5 mm. After locating the desired level of the colon by colonoscopy, they created a window at the mesocolon, passed the sheet circumferentially, and sutured its edges in order to produce the stenosis and observed the animals for 14-19 days. Putranto et al. [140] utilized three plastic cable ties placed at a 1 cm distance over an encircled standard plastic sheet covering the intestines and reported effective fibrosis in 24 h postoperatively when animals were euthanized. Lukas et al. [141] developed a model of surgically induced stricture to mimic CD anastomotic strictures. In this method, authors performed a modified Roux-en-Y side-to-side ileosigmoid anastomosis followed by at least four postoperative endoscopic injections of phenol/trinitrobenzenesulfonic acid solution at the anastomotic site; 15 of the 19 pigs used completed the 6 months follow up and only one adverse event of rectal prolapse is documented.

The main question raised in partial IO models is how long these animals survive with these techniques and how some of these partial obstruction models are different from chronic complete obstruction models, especially when the technique is about forming a ring-like structure around the intestinal lumen. Sun et al. [124, 142] commented that their 20 x 10 mm rectangular polyethylene ring placed on the jejunum of male rats to induce partial IO was looser than the 30 x 5 mm rectangular polyethylene ring on the ileum of male rats for complete IO, yet this conclusions lack proved consistency. Galvez et al. [121] utilized three techniques of experimental partial IO in rats: a longitudinal serosal approximation of the intestinal lumen estimated to cause 50% partial, a 5-mm-wide GoreTex ring around the colon estimated to cause an obstruction of approximately 35% of the intestinal lumen and a suture fixated to an avascular section of the mesentery before encircling the colon and tied up to a 15% partial IO. Of the three study groups, none of the gore-text ring subjects survived the cut-off of 6 weeks. Georgopoulos et al. [9], in an experimental comparative study of all techniques for partial IO, comment that all rings failed to meet the 4-week survival cut-off. Two weeks is considered the least time needed for histopathological characteristics of partial obstruction to develop [9, 75]. Thus, if a chronic partial intestinal obstruction such as stenosis is the aim of the study, the experimental model needs to establish a long-standing obstruction with survival rates longer than 2 weeks.

*Adhesion models:* Intraperitoneal adhesions can result in both complete and partial IO. As surgical instruments and materials are still evolving, experimental adhesion models are an ongoing field of research. All operative aspects may act as adhesiogenic factors by inducing a foreign-body type reaction. Mechanical and chemical intraoperative stimuli, such as types of gloves, types of gauze, moisture level of gauze, forceps, intraperitoneal anesthetic agents, or lavage, all induce inflammatory reactions to the intestinal wall and need to be carefully and in detail addressed in a protocol of experimental adhesive IO. Cecal abrasion alone or combined with other methods of intestinal manipulation is probably the most commonly cited method for postoperative intestinal adhesions, and it is accepted that the extent of adhesion formation is directly proportional to the extent of abrasion, even though quality and quantity are not standardized [23]. Simple laparotomy or laparoscopy with or without handling of the intestine, abrasion of cecum with or without abrasion of other organs, abrasion of the peritoneum, abrasion of the abdominal wall, intraperitoneal injection of various substances, drains, and modification of these experimental techniques by changing force, time, suturing materials, gloves, use of electrocautery, healing agents, meshes, inflammation agents and instrumentation have all been applied to replicate postoperative adhesive bowel obstruction [23, 143–153]. Authors have also published their experiments on how radiation, desiccation, thermal injury, bleeding, ischemia, endometriosis, cancer, pain, and peritoneal irritation affect the formation of adhesions [23, 154–158]. The main drawback of all of these techniques is the inability to standardize them in an objective countable method.

### 3.2 Functional intestinal obstruction

This type of IO refers to a wide spectrum of diseases characterized by intestinal transit alterations that mimic the signs and symptoms of mechanical IO as described in the introduction, yet there is no actual anatomic barrier. Functional IO represents an asynergia of the muscular and nervous intestinal plexus due to either a disruption of their homeostasis, genetic alterations, or because of a systematic disease. Also described as paralysis of the intestine or ileus, functional IO may affect both the small and the large intestine, whereas pseudo-obstruction is a functional obstruction affecting only the colon. Most common causes are postoperative ileus after any kind of surgery, intrabdominal infections, sepsis, decreased blood supply, pharmaceutical side effects, cystic fibrosis, neuromuscular diseases, and prolonged hospitalization [55, 159, 160]. Transgenic, knock-out and genetically modified animal models are available and commonly used to study functional IO. The surgical models of functional IO aim to induce postoperative ileus, intestinal ischemia, or intraperitoneal infections.

*Postoperative ileus models:* A common technique to replicate postoperative paralysis is laparotomy with the handling of the intestine by translocation, eventration, hand pressure, or exposure to room air for a period of time [161–170]. This postoperative ileus model is cost-effective and not demanding by means of surgical skills, yet it is highly variable as pressure is not countable, and neither is the handling of the intestine. Authors utilize different time intervals of what they record as intestine manipulation (**Table 6**), and comparative studies do not exist regarding time intervals of tissue handling and means of manipulation. Gerring et al. [171], after exposure of

| Author | Method | Time (minutes) | Species | Adverse events | Sampling |
|---|---|---|---|---|---|
| Schwarz [161] | Eventration and placement on moist gauze, then light manipulation of the entire bowel with two moist cotton applicators | 15' | Male rats | None | 0 h, 24 h, 3d |
| Yong-Yu Li [162] | Eventration and manipulation with two moist cotton applicators | 5' | Female mice | n/a | 24 h |
| Morris [163] | Room air exposure | 30′ | Labrador dogs | n/a | 1-22d |
| Vilz [164] | Eventration, placement on moist gauze, and manual evacuation by rolling two moist and sterile cotton applicators | n/a | Male mice | Caution for hemorrhage | 24 hr |
| Hartmann [165] | Manipulation with two moist cotton applicators ±mechanical bowel preparation or selective decontamination | 15 min | Mice | n/a | 1 h, 3 h,9 h |
| The [166] | Exteriorization and manipulation with sterile, moist cotton applicators | 5′ | Female mice | n/a | 24 hr |
| De Winter [167] | Evisceration on sterile gauze and manipulation by hand pressure | 5′ | Male rats | n/a | 20 minutes |
| De Winter [168] | Evisceration on sterile gauze and manipulation (not defined) | 5' | Male rats | n/a | 20 minutes |
| Matsumoto [169] | Exposure and gentle manipulation with a sterile moistened cotton swab | 3–5' | Male mice | n/a | n/a |
| Sun [170] | Evisceration on wet gauze and manipulation with a moist medical cotton swab three times | 10′ | Male mice | None | 3 h, 24 h |
| Gerring [171] | Evisceration and vigorous rub with dry swab, then covered with dry towels | 40' | ponies | none | 3 h to 74 h |
| Van Bre [172] | Evisceration with wet gauze<br>1. Three times roll on Plexiglas platform with cotton applicator device<br>2. Compression with moist cotton applicators | n/a | mice | n/a | 24 h |

**Table 6.**
*Variations of commonly used techniques of experimental postoperative ileus.*

the small intestine, used a dry swab to rub it vigorously for 10 minutes and then covered it with a dry towel for a further 30 minutes. Van Bree et al. [172] in order to eliminate pressure and handling variations, used a plexiglass onto which they rolled the intestine three to four times with the use of a cotton applicator and reported that they managed to reduce variability in outcomes; thus, they can reduce the number of animals used. Many authors record the intestine manipulation technique not as the study group but as the sham operation to compare the outcomes of intestinal motility, inflammation, management, and recovery compared to intervention groups.

*Peritoneal irritation models:* Functional ileus because of chemical or inflammatory origin is easily induced by injecting a substance intraperitoneally, subcutaneously, intragastrically, intravenously, intranasally, or inhaled. This is a feasible way to replicate both septic, systematic (such as diabetic), and pharmaceutical functional ileus, and there is a long-listed catalog of agents used depending on the aim of each protocol. Lipopolysaccharides, acetic acid, iodine, capsaicin, bacteria, viruses, chemotherapeutic agents, anesthetic drugs, and multiple other substances have been administered to animal models with subsequent study of the enteric transit time and pathophysiological reactions [173–181]. Some authors utilize the two-hit technique by sequential administration of two inflammatory stimuli. The cecal ligation and puncture technique is another commonly used technique to induce septic paralysis. The cecum is ligated directly beneath the ileocecal valve, generating an inflammatory response resulting in necrotic tissue. This step is followed by perforating the cecum, facilitating the leakage of fecal material into the typically sterile peritoneal cavity. As a consequence, animals exhibit classical symptoms of sepsis and typically succumb to the condition. [182]. These are mainly surgical models used to study the effect of sepsis, but the gut is also an affected organ in septic conditions and has been used to study septic ileus by Overhaus et al. and Köylüoğlu et al. [183, 184].

*Ischemia-Reperfusion models:* The intestine is one of the most vulnerable organs to ischemia-reperfusion injuries [185]. A common animal model of intestinal reperfusion injury is occlusion of the superior mesenteric artery (SMA) by atraumatic microvascular clamps or by ligation in rodents for varying time intervals. Heparin is often administered intravenously to prevent thrombus formation within the SMA, enabling the re-establishment of circulation once the clamps are removed. The outcome depends on the selected part of the intestine as perfusion differs along the GIT, with the jejunum, ileum, and large colon showing different levels of resistance [100]. SMA occlusion alone produces variable injury severity and results in high mortality rates. When SMA occlusion is combined with ligation of collateral arcades, a more consistent injury is induced with lower mortality levels. The low-flow ischemia model uses adjustable atraumatic clamps to lower SMA flow at 20% of baseline levels to replicate hypovolemic clinical scenarios and is reported mainly in cats. Segmental mesenteric vascular occlusion by clamping the local mesenteric vascular supply and cross-clamping the bowel is a strangulation model for intestinal reperfusion injury as described above. A more recent technique in porcine models involves embolization of the SMA using materials such as buthyl-2-cyanoacrylate or polyvinyl alcohol particles and gel foam. This approach allows for complete, irreversible ischemia without exposing or manipulating the abdominal contents, making it suitable for studies of acute mesenteric ischemia [100].

## 4. Conclusions

The basic principles of any IO experimental model are:

1. Selection of the proper animal species based on research objectives, availability, ethical considerations, and translational relevance to human physiology.

2. Adhere to ethical guidelines and regulations governing animal research, ensuring humane treatment of animal subjects and minimizing pain and distress associated with experimental procedures.

3. Plan a reproducible and consistent protocol for experimental intestinal obstruction.

4. Introduce a mechanical or functional obstruction in the intestine of the animal model.

5. Regularly monitor the animal model for signs of intestinal obstruction, including changes in behavior, food intake, abdominal distension, and bowel movements, and monitor for potential complications such as intestinal ischemia, perforation, and sepsis.

6. Conduct histological analysis of intestinal tissue samples to evaluate morphological changes associated with obstruction, such as mucosal damage, inflammation, and fibrosis, and perform molecular analyses to investigate gene expression patterns, signaling pathways, and molecular mechanisms involved in the development and progression of intestinal obstruction.

7. Consider conducting longitudinal studies to assess the progression of intestinal obstruction over time and evaluate the long-term effects on gastrointestinal function, mucosal integrity, and overall health.

## Author details

Eleftheria Mavrigiannaki[1,2]* and Ioannis Georgopoulos[1]

1 Children's Hospital "Agia Sofia", Athens, Greece

2 Mitera Children's Hospital, Greece

*Address all correspondence to: ele17mav@yahoo.gr

## References

[1] Askitopoulou H, Stefanakis G, Astyrakaki EE, Papaioannou A, Agouridakis P. Emergencies and acute diseases in the collected works of hippocrates: Observation, examination, prognosis, therapy. European Journal of Emergency Medicine. [Internet]. 2016;**23**(6):399-405. Available from: https://www.pubmed.ncbi.nlm.nih.gov/27384218/

[2] Ballantyne GH. The meaning of ileus. Its changing definition over three millennia. American Journal of Surgery. [Internet]. 1984;**148**(2):252-256. Available from: https://www.pubmed.ncbi.nlm.nih.gov/6380325/

[3] Cappell MS, Batke M. Mechanical obstruction of the small bowel and colon. The Medical Clinics of North America. 2008;**92**(3):575-597

[4] Rami, Reddy SR, Cappell MS. A systematic review of the clinical presentation, diagnosis, and treatment of small bowel obstruction. Current Gastroenterology Reports. 2017;**19**(6)

[5] Chapple KS, Hartley JE. Intestinal obstruction. British Journal of Hospital Medicine. 2006;**67**(Sup1):M5-M7. DOI: 10.12968/hmed.2006.67.sup1.2

[6] Shelton BK. Intestinal obstruction. AACN (American Association of Critical-Care Nurses) Advanced Practice in Acute Critical Care. 1999;**10**(4):478-491. Available from: https://www.aacnjournals.org/aacnacconline/article-abstract/10/4/478/13809/Intestinal-Obstruction?redirectedFrom=PDF

[7] Long D, Mao C, Liu Y, Zhou T, Xu Y, Zhu Y. Global, regional, and national burden of intestinal obstruction from 1990 to 2019: An analysis from the global burden of disease study 2019. International Journal of Colorectal Disease. 2023;**38**(1):1-16. [Internet]. Available from: DOI: 10.1007/s00384-023-04522-6

[8] Georgopoulos I, Mavrigiannaki E, Stasinopoulou S, Renieris G, Nikolakis G, Bamias G, et al. Experimental intestinal stenosis alters crohn's disease-like intestinal inflammation in ileitis-prone mice. Digestive Diseases and Sciences. [Internet]. 2022;**67**(5):1783-1793. Available from: https://www.pubmed.ncbi.nlm.nih.gov/34350516/

[9] Georgopoulos I, Mavrigiannaki E, Stasinopoulou S, Renieris G, Nikolakis G, Chaniotakis I, et al. Experimental models of partial intestinal obstruction in young mice, establishment, and evaluation. The Journal of Surgical Research. [Internet]. 2020;**252**:206-215. Available from: http://www.journalofsurgicalresearch.com/article/S0022480420301232/fulltext

[10] Banstola A, Reynolds JNJ. The sheep as a large animal model for the investigation and treatment of human disorders. Biology (Basel). 2022;**11**(9):1-26

[11] Richmond J. The three Rs: A journey or a destination? Alternatives to Laboratory Animals. 2000;**28**(6):761-773

[12] Hubrecht R, Kirkwood J. The UFAW Handbook on the Care and Management of Laboratory and Other Research Animals. 8th ed. UFAW Handb Care Manag Lab Other Res Anim. UK: Wiley-Blackwell, Wiley European Distribution Centre; 2010

[13] Russell WMS, Burch RL. Russell and Burch's Principles of Humane Experimental Techniques. Special

Edition. Wheathampstead, UK: UFAW; 1992. 238 p

[14] ARCHER RK. Alternatives to Animal Experiments. By Professor D. H. Smyth. Equine Veterinary Journal, 1978;**10**(4):270. DOI: 10.1111/j.2042-3306.1978.tb02282.x

[15] Guide for the Care and Use of Laboratory Animals. National Research Council (U.S.). Committee for the Update of the Guide for the Care and Use of Laboratory Animals. Institute for Laboratory Animal Research (U.S.). NW Washington, DC: The National Academies Press; [Internet] 2011. Available from: https://www.ncbi.nlm.nih.gov/books/NBK54050/

[16] National Centre for the Refinement & Reduction of Animals in Research., "The 3Rs". 2017. https://www.nc3rs.org.uk/who-we-are/3rs [Accessed: July 01, 2024]

[17] Kilkenny C, Parsons N, Kadyszewski E, Festing MFW, Cuthill IC, Fry D, et al. Survey of the quality of experimental design, statistical analysis and reporting of research using animals. PLoS One. [Internet]. 2009;**4**(11). Available from: https://www.pubmed.ncbi.nlm.nih.gov/19956596/

[18] Kilkenny C, Browne WJ, Cuthill IC, Emerson M, Altman DG. Improving bioscience research reporting: The arrive guidelines for reporting animal research. PLoS Biology. [Internet]. 2010;**8**(6). Available from: https://www.pubmed.ncbi.nlm.nih.gov/20613859/

[19] du Sert NP, Ahluwalia A, Alam S, Avey MT, Baker M, Browne WJ, et al. Reporting animal research: Explanation and elaboration for the ARRIVE guidelines 2.0. PLoS Biology. [Internet]. 2020;**18**(7). Available from: https://www.pubmed.ncbi.nlm.nih.gov/32663221/

[20] The European Parliament and the Council of the European Union. Directive 2010/63/EU of the European Paliament and of the Council of 22 September 2010 on the protection of animals used for scientific purposes. Official Journal of the European Union. 2010;**276**:33-79

[21] Republic P of the G. Presidential Decree 56/2013. Government Gazette. Athens, P.K: National Printing House of Greece; 2013. p. 53

[22] Tannenbaum J, Bennett BT. Russell and Burch's 3Rs then and now: The need for clarity in definition and purpose. Journal of the American Association for Laboratory Animal Science. [Internet]. 2015;**54**(2):120. Available from: /pmc/articles/PMC4382615/

[23] Wiseman DM. Animal adhesion models: Design, variables, and relevance. Peritoneal Surgery. [Internet]. 2000:459-476. Available from: https://www.link.springer.com/chapter/10.1007/978-1-4612-1194-5_38

[24] Hatton GB, Yadav V, Basit AW, Merchant HA. Animal farm: Considerations in animal gastrointestinal physiology and relevance to drug delivery in humans. Journal of Pharmaceutical Sciences. [Internet]. 2015;**104**(9):2747-2776. Available from: http://www.jpharmsci.org/article/S0022354916300569/fulltext

[25] Oliveira APL, Rangel JPP, Riodades LFS, Almeida BL, Mathias CHT, Conti LMC, et al. Establishment of an experimental model of small intestinal ischemia and reperfusion injuries in New Zealand rabbits. Pesquisa Veterinária Brasileira. [Internet]. 2018;**38**(8):1664-1674. Available from: https://www.scielo.br/j/pvb/a/hJLXPYRKfhDGyDbrW9NdWfx/?lang=en

[26] South FL, Hardt LLJ. Experimental intestinal obstruction. Archives of Internal Medicine. 1918;**XXI**(2):292-308

[27] Xilong L, Lei S. Changes of Tissue Endothelin-1 and Nitric Oxide Synthase in a Sheep Model of Large Intestinal Obstruction. Veterinary Research Communications. **28**(8):719-725

[28] Gingerich DA, Murdick PW. Experimentally induced intestinal obstruction in sheep: Paradoxical aciduria in metabolic alkalosis. American Journal of Veterinary Research. 1975;**36**(5):663-668

[29] Weatherall D. The use of non-human primates in research. The Academy of Medical Sciences. 2006;**153**

[30] Chatfield K, Morton D. The Use of Non-human Primates in Research. Cham: Springer; 2018. pp. 81-90. Available from: https://www.link.springer.com/chapter/10.1007/978-3-319-64731-9_10 Print ISBN: 978-3-319-64730-2, Online ISBN: 978-3-319-64731-9

[31] Calasans-Maia MD, Monteiro ML, Áscoli FO, Granjeiro JM. The rabbit as an animal model for experimental surgery. Acta Cirúrgica Brasileira. [Internet]. 2009;**24**(4):325-328. Available from: https://www.scielo.br/j/acb/a/7JyTfL678qtnXSgMTsnWRDp/

[32] Frohlich J. Rats and mice. [Internet]. In: Ferrets, Rabbits, and Rodents: Clinical Medicine and Surgery. 4th ed. St. Louis, Missouri: Elsevier; 2020. 345-367 p. DOI: 10.1016/B978-0-323-48435-0.00025-3

[33] Jiminez JA, Uwiera TC, Inglis GD, Uwiera RRE. Animal models to study acute and chronic intestinal inflammation in mammals. Gut Pathogens. [Internet]. 2015;**7**(1):29. Available from: /pmc/articles/PMC4641401/

[34] Tohyama S, Kobayashi E. Age-appropriateness of porcine models used for cell transplantation. Cell Transplantation. [Internet]. 2019;**28**(2):224-228. Available from: /pmc/articles/PMC6362526/

[35] Walters EM, Prather RS. Advancing swine models for human health and diseases. Missouri Medicine. [Internet]. 2013;**110**(3):212-215. Available from: http://www.ncbi.nlm.nih.gov/pubmed/6179855

[36] Lee JG, Sung YH, Baek IJ. Generation of genetically-engineered animals using engineered endonucleases. Archives of Pharmacal Research. [Internet]. 2018;**41**(9):885-897. DOI: 10.1007/s12272-018-1037-z

[37] Kararli TT. Comparison of the gastrointestinal anatomy, physiology, and biochemistry of humans and commonly used laboratory animals. Biopharmaceutics & Drug Disposition. [Internet]. 1995;**16**(5):351-380. DOI: 10.1002/bdd.2510160502

[38] de Zwart LL, Rompelberg CJM, Sips AJAM, Welink J, van Engelen JGM. Anatomical and physiological differences betweenvarious species used in studies on the pharmacokinetics and toxicology of xenobiotics. A review of literature. BA Bilthoven: Rijksinstituut voor Volksgezondheid en Milieu RIVM; 1999. Available from: https://www.rivm.openrepository.com/handle/10029/10210

[39] Johnson-Delaney C. Anatomy and Physiology of the Rabbit and Rodent Gastrointestinal System. USA: Association of Avian Veterinarians, From EastsideAvian & ExoticAnimal Medical Center, PLLC; 2006. Available

from: https://www.researchgate.net/publication/237308548

[40] Ziegler A, Gonzalez L, Blikslager A. Large animal models: The key to translational discovery in digestive disease research. Cellular and Molecular Gastroenterology and Hepatology. [Internet]. 2016;**2**(6):716. Available from: /pmc/articles/PMC5235339/

[41] Schaaf CR, Gonzalez LM. Use of translational, genetically modified porcine models to ultimately improve intestinal disease treatment. Frontiers in Veterinary Science. [Internet]. 2022;**9**. Available from: https://www.pubmed.ncbi.nlm.nih.gov/35669174/

[42] Casteleyn C, Rekecki A, Van Der Aa A, Simoens P, Van Den Broeck W. Surface area assessment of the murine intestinal tract as a prerequisite for oral dose translation from mouse to man. Laboratory Animals. 2010;**44**(3):176-183

[43] Nguyen TLA, Vieira-Silva S, Liston A, Raes J. How informative is the mouse for human gut microbiota research? Disease Models & Mechanisms. [Internet]. 2015;**8**(1):1-16. Available from: https://www.pubmed.ncbi.nlm.nih.gov/25561744/

[44] Bennett BT, Brown MJ, Schofield JC. Essentials for Animal Research a Primer for Research Personnel Provided by the Animal Welfare Information Center. 2nd ed. Beltsville, Md: United States Department of Agriculture National Agricultural Library; 1995. Available from: http://www.nal.usda.gov/awic/pubs/noawicpubs/essentia.htm

[45] American Veterinary Medical Association. AVMA Guidelines for the Euthanasia of Animals. 2013 ed. Schaumburg, IL. 2013

[46] Berlin SC, Goske MJ, Obuchowski N, Alexander F, Zepp RC, Goldblum JR, et al. Small bowel obstruction in rats: Diagnostic accuracy of sonography versus radiography. Journal of Ultrasound in Medicine. [Internet]. 1998;**17**(8):497-504. DOI: 10.7863/jum.1998.17.8.497

[47] Fish R, Danneman P, Brown M, Karas A. Anesthesia and Analgesia in Laboratory Animals. Elsevier Inc.; 2008. ISBN 978-0-12-373898-1

[48] Cicero L, Fazzotta S, Palumbo VD, Cassata G, Lo Monte AI. Anesthesia protocols in laboratory animals used for scientific purposes. Acta Biomed. 8 Oct 2018;**89**(3):337-342. DOI: 10.23750/abm.v89i3.5824. PMID: 30333456; PMCID: PMC6502126

[49] Richardson CA, Flecknell PA. Anaesthesia and post-operative analgesia following experimental surgery in laboratory rodents: Are we making progress? Alternatives to Laboratory Animals. [Internet]. 2005;**33**(2):119-127. Available from: https://www.pubmed.ncbi.nlm.nih.gov/16180987/

[50] National Research Council of the National Academies. Guide for the care and use of laboratory animals. Washington, DC: National Academies Press; 2011. Available from: http://www.nap.edu

[51] Otto K, von Thaden AK. Anaesthesia, analgesia and euthanasia. The Laboratory Rat. Chapter 5.4. Elsevier Ltd.; 2012:739-759. ISBN 978-0-12-382008-2

[52] Chaniotakis I, Spyrliadis A, Katsimpoulas M, Kostomitsopoulos N. The mouse and the rat in surgical research. The anesthetic approach. Journal of the Hellenic Veterinary Medical Society. 2016;**67**(3)

[53] Institutional Animal Care & Use Committee Office. Tulane Office of

Research. [Internet]. Available from: https://www.research.tulane.edu/iacuc

[54] Bordeianou L, Yeh DD. Etiologies, clinical manifestations, and diagnosis of mechanical small bowel obstruction in adults. UpToDate. Literature review current through. 2022:1-69. Available from: https://www-uptodate-com.pbidi.unam.mx:2443/contents/etiologies-clinical-manifestations-and-diagnosis-of-mechanical-small-bowel-obstruction-in-adults?search=oclusiónintestinal&source=search_result&selectedTitle=1~150&usage_type=default&display_rank=1#H16

[55] Griffiths S, Glancy DG. Intestinal obstruction. Surgery (United Kingdom). [Internet]. 2020;**38**(1):43-50. Available from: http://www.surgeryjournal.co.uk/article/S0263931919302200/fulltext

[56] Wu CC, Lu YZ, Wu LL, Yu LC. Role of myosin light chain kinase in intestinal epithelial barrier defects in a rat model of bowel obstruction. BMC Gastroenterology. [Internet]. 2010;**10**:39. Available from: /pmc/articles/PMC2868795/

[57] Şen V, Uluca Ü, Ece A, Güneş A, Zeytun H, Arslan S, et al. Role of Ankaferd on bacterial translocation and inflammatory response in an experimental rat model of intestinal obstruction. International Journal of Clinical and Experimental Medicine. [Internet]. 2014;**7**(9):2677. Available from: /pmc/articles/PMC4211775/

[58] Deitch EA, Bridges WM, Ma JW, Ma L, Berg RD, Specian RD. Obstructed intestine as a reservoir for systemic infection. American Journal of Surgery. [Internet]. 1990;**159**(4):394-401. Available from: https://www.pubmed.ncbi.nlm.nih.gov/2316803/

[59] Pittner A, Nalos M, Theisen M, Ploner F, Brückner UB, Georgieff M, et al. Inhaling nitrous oxide or xenon does not influence bowel wall energy balance during porcine bowel obstruction. Anesthesia and Analgesia. [Internet]. 2002;**94**(6):1510-1516. Available from: https://www.journals.lww.com/anesthesia-analgesia/fulltext/2002/06000/inhaling_nitrous_oxide_or_xenon_does_not_influence.25.aspx

[60] Aldemir M, Kökoğlu ÖF, Geyik MF, Büyükbayram H. Effects of octreotide acetate and Saccharomyces boulardii on bacterial translocation in an experimental intestinal loop obstruction model of rats. The Tohoku Journal of Experimental Medicine. [Internet]. 2002;**198**(1):1-9. Available from: https://www.pubmed.ncbi.nlm.nih.gov/12498309/

[61] Akyildiz M, Ersin S, Oymaci E, Dayangac M, Kapkac M. Effects of somatostatin analogues and vitamin C on bacterial translocation in an experimental intestinal obstruction model of rats. Journal of Investigative Surgery. 2000;**13**(3):169-173

[62] Shields R. The absorption and secretion of fluid and electrolytes by the obstructed bowel. The British Journal of Surgery. 1965;**52**(10):774-779

[63] Yildiz H, Oncel M, Kurt N, Vural S, Gezen C, Dalkilic G, et al. The relationship between the level of thirteen different substances and enzymes in blood and peritoneal fluid and the duration of mechanical intestinal obstruction: An experimental study on rats. Ulusal Travma ve Acil Cerrahi Dergisi. 2003;**9**(3):183-188

[64] Chang TM, Lu RH, Tsai LM. Glutamine ameliorates mechanical obstruction-induced intestinal injury. The Journal of Surgical Research. 2001;**95**(2):133-140

[65] Edizsoy A, Yılmaz E, Çevikel MH, Yenisey Ç, Sakarya S, Meteoğlu İ. Gut mucosa in the rats exposed temporary mechanical obstruction fed with probiotic. Ulusal Travma ve Acil Cerrahi Dergisi. 2020;**26**(6):833-842

[66] Mueller MH, Zhao X, Macheroux T, Kasparek MS, Seeliger H, Kreis ME. Differential activation of afferent neuronal and inflammatory pathways during small bowel obstruction (SBO). Neurogastroenterology and Motility. 2016;**28**(10):1599-1608

[67] Whipple GH, Stone HB, Bernheim BM. Intestinal obstruction: I. A study of a toxic substance produced in closed duodenal loops. The Journal of Experimental Medicine. [Internet]. 1913;**17**(3):286-306. Available from: https://www.pubmed.ncbi.nlm.nih.gov/19867644/

[68] Dragstedt LR, Moorhead JJ, Burcky FW. Intestinal obstruction: An experimental study of the intoxication in closed intestinal loops. The Journal of Experimental Medicine. [Internet]. 1917;**25**(3):421-439. Available from: https://www.pubmed.ncbi.nlm.nih.gov/19868099/

[69] Stone HB, Bernheim BM, Whipple GH III. The experimental study of intestinal obstruction. Annals of Surgery. [Internet]. 1914;**59**(5):714. Available from: https://www.ncbi.nlm.nih.gov/pmc/articles/PMC1406381/

[70] Kabaroudis A, Papaziogas B, Koutelidakis I, Kyparissi-Kanellaki M, Kouzi-Koliakou K, Papaziogas T. Disruption of the small-intestine mucosal barrier after intestinal occlusion: A study with light and electron microscopy. Journal of Investigative Surgery. 2003;**16**(1):23-28

[71] Mulvihill SJ, Pappas TN, Fonkalsrud EW, Debas HT. The effect of somatostatin on experimental intestinal obstruction. Annals of Surgery. [Internet]. 1988;**207**(2):169. Available from: /pmc/articles/PMC1493368/?report=abstract

[72] Firat U, Senol S, Gelincik I, Kapan M, Tokgoz O, Tekin R, et al. The effects of caffeic acid phenethyl ester (CAPE) on bacterial translocation and inflammatory response in an experimental intestinal obstruction model in rats. European Review for Medical and Pharmacological Sciences. 2015;**19**:1907-1914

[73] Hartmann L, Zhao X, Macheroux T, Kasparek MS, Kreis ME, Mueller MH. Time-dependent alterations of gut wall integrity in small bowel obstruction in mice. The Journal of Surgical Research. [Internet]. 2019;**233**:249-255. Available from: https://www.pubmed.ncbi.nlm.nih.gov/30502255/

[74] Enochsson L, Hellström PM, Nylander G, Johansson C. Myoelectric motility patterns during mechanical obstruction and paralysis of the small intestine in the rat. Scandinavian Journal of Gastroenterology. [Internet]. 1987;**22**(8):969-974. DOI: 10.3109/00365528708991944

[75] Youb CI, Glasgow NJ, Takayama I, Horiguchi K, Sanders KM, Ward SM. Loss of interstitial cells of Cajal and development of electrical dysfunction in murine small bowel obstruction. The Journal of Physiology. 2004;**536**(2):555-568

[76] Storkholm JH, Zhao J, Villadsen GE, Hager H, Jensen SL, Gregersen H. Biomechanical remodeling of the chronically obstructed guinea pig small intestine. Digestive Diseases and Sciences. [Internet]. 2007;**52**(2):336-346. DOI: 10.1007/s10620-006-9431-7

[77] Barros Paula Nunes BLB, Saad SS, Jucá MJ, Porfírio Z, Matos D. Analysis

of bacteremia occurring in the presence of obstruction of the left colon in rats submitted to transoperative antegrade mechanical lavage. Journal of Investigative Surgery. 2005;**18**(5):233-240

[78] Graeber GM, O'Neill JF, Wolf RE, Wukich DK, Cafferty PJ, Harmon JW. Elevated levels of peripheral serum creatine phosphokinase with strangulated small bowel obstruction. Archives of Surgery. 1983;**118**(7):837-840

[79] Amanova DY, Lavrinenko AV, Kaliyeva DK, Matyushko DN, Ivachyov PA, Turgunov YM. Comparative evaluation of translocation of GFP producing *Escherichia coli* strains in acute intestinal obstruction. Bulletin of Experimental Biology and Medicine. 2019;**167**(5):660-662

[80] Zhang N, Ma JH, Qi M, Zhou Z Li. Effects of changes in the intestinal function of rats with intestinal dysfunction induced by new intestinal obstructions. International Journal of Clinical and Experimental Medicine. 2018;**11**(11):11631-11642

[81] Li L, Zou C, Zhou Z, Wang X, Yu X. Phenotypic changes of interstitial cells of Cajal after intestinal obstruction in rat model. Brazilian Journal of Medical and Biological Research. [Internet]. 2019;**52**(10). Available from: /pmc/articles/PMC6799941/

[82] Yu XY, Zou CL, Zhou ZL, Shan T, Li DH, Cui NQ. Phasic study of intestinal homeostasis disruption in experimental intestinal obstruction. World Journal of Gastroenterology. [Internet]. 2014;**20**(25):8130-8138. Available from: https://www.wjgnet.com/1007-9327/full/v20/i25/WJG-20-8130-g004.htm

[83] Mo J, Gao L, Zhang N, Xie J, Li D, Shan T, et al. Structural and quantitative alterations of gut microbiota in experimental small bowel obstruction. PLoS One. [Internet]. 2021;**16**(8):e0255651. Available from: https://www.journals.plos.org/plosone/article?id=10.1371/journal.pone.0255651

[84] Mo J, Zhang N, Li D, Fan L, Xie J. Reversible small bowel obstruction in rats. International Journal of Clinical and Experimental Medicine. 2020;**13**(4):2276-2285

[85] Correa-Martin L, Parraga E, Sanchez-Margallo FM, Latorre R, Lopez-Albors O, Wise R, et al. Mechanical intestinal obstruction in a porcine model: Effects of intra-abdominal hypertension. A preliminary study. PLoS One. [Internet]. 2016;**11**(2):e0148058. Available from: https://www.journals.plos.org/plosone/article?id=10.1371/journal.pone.0148058

[86] Ros EP, Correa-Martín L, Sánchez-Margallo FM, Candanosa-Aranda IE, Malbrain MLNG, Wise R, et al. Time-course evaluation of intestinal structural disorders in a porcine model of intra-abdominal hypertension by mechanical intestinal obstruction. PLoS One. [Internet]. 2018;**13**(1). Available from: /pmc/articles/PMC5777654/

[87] Kaszaki J, Palásthy Z, Érczes D, Rácz A, Torday C, Varga G, et al. Kynurenic acid inhibits intestinal hypermotility and xanthine oxidase activity during experimental colon obstruction in dogs. Neurogastroenterology and Motility. 2008;**20**(1):53-62

[88] Roscher R, Oettinger W, Beger HG. Bacterial microflora, endogenous endotoxin, and prostaglandins in small bowel obstruction. American Journal of Surgery. 1988;**155**(2):348-355

[89] Fraser I. Motility changes associated with large bowel obstruction and its surgical relief. Annals of the Royal College of Surgeons of England. [Internet]. 1984;**66**(5):321. Available from: /pmc/articles/PMC2493677/?report=abstract

[90] Fraser ID, Condon RE, Schulte WJ, Decosse JJ, Cowles VE. A device to simulate the mechanical component of malignant bowel obstruction. Archives of Surgery. [Internet]. 1981;**116**(2):194-196. Available from: https://www.jamanetwork.com/journals/jamasurgery/fullarticle/587135

[91] Oruç MT, Özmen MM, Kazan O, Düzgün AP, Özkara HA, Arik D, et al. Does serum hexosaminidase activity play a role in the diagnosis of strangulated bowel obstruction an experimental study. Digestive Diseases and Sciences. 2004;**49**(10):1681-1686

[92] Inoue A, Nitta N, Ota S, Takaki K, Imai Y, Misaki S, et al. MR imaging-based evaluation of mesenteric ischemia caused by strangulated small bowel obstruction and mesenteric venous occlusion: An experimental study using rabbits. Magnetic Resonance in Medical Sciences. 2020;**19**(2):125-134

[93] Zanoni FL, Benabou S, Greco KV, Moreno ACR, Costa Cruz JWM, Filgueira FP, et al. Mesenteric microcirculatory dysfunctions and translocation of indigenous bacteria in a rat model of strangulated small bowel obstruction. Clinics. 2009;**64**(9):911-919

[94] El-Awady SI, El-Nagar M, El-Dakar M, Ragab M, Elnady G. Bacterial translocation in an experimental intestinal obstruction model. C-reactive protein reliability? Acta Cir Bras. Mar-Apr; 2009;**24**(2):98-106. DOI: 10.1590/s0102-86502009000200005. PMID: 19377777

[95] Bornside GH, Cohn I. Clostridial toxins in strangulation intestinal obstruction in the rabbit. Annals of Surgery. 1960;**152**(1):330-342

[96] Samel S, Keese M, Kleczka M, Lanig S, Gretz N, Hafner M, et al. Microscopy of bacterial translocation during small bowel obstruction andischemia in vivo - A new animal model. BMC Surgery. 2002;**2**:1-7

[97] Yani A, Dorothy D, Amaliah R. Influence of intestinal strangulation release on ischemiareperfusion injury in sprague dawley rats. Annals of African Surgery. [Internet]. 2021;**18**(2):90-95. Available from: https://www.scholar.ui.ac.id/en/publications/influence-of-intestinal-strangulation-release-on-ischemiareperfus

[98] Cakir M, Yildirim D, Sarac F, Donmez T, Mirapoglu S, Hut A, et al. In the experimental model of acute mesenteric ischemia, the correlation of blood diagnostic parameters with the duration of ischemia and their effects on choice of treatment. Journal of Investigative Surgery. [Internet]. 2019;**32**(6):507-514. DOI: 10.1080/08941939.2018.1437486

[99] Chen Y, Qin C, Wang G, Xiao M, Xiao G, Luan Z, et al. Contribution of heparin to recovery of incarcerated intestine in a rat incarcerated hernia model. Hernia. [Internet]. 2019;**23**(6):1155-1161. DOI: 10.1007/s10029-019-01985-x

[100] Gonzalez LM, Moeser AJ, Blikslager AT. Animal models of ischemia-reperfusion-induced intestinal injury: Progress and promise for translational research. American Journal of Physiology. Gastrointestinal and Liver Physiology. 2015;**308**(2):G63-G75

[101] Youssef HA, Ali MM, Kuraa HMM. Effect of small intestine

strangulation obstruction on clinical and histopathological parameters an experimental study in donkeys. Veterinary Research Forum. [Internet]. 2011;**2**(2):75-86. Available from: http://www.doaj.org/doaj?func=fulltext&am p;passMe=http://www.urmia.ac.ir/vrf/SharedDocuments/pdf/vol-2no-2/75.pdf

[102] Rasslan R, Utiyama EM, Marques GMN, Ferreira TC, da Costa VAP, de Victo NC, et al. Inflammatory activity modulation by hypertonic saline and pentoxifylline in a rat model of strangulated closed loop small bowel obstruction. International Journal of Surgery. Veterinary Research Forum. [Internet]. 2014;**12**(6):594-600. DOI: 10.1016/j.ijsu.2014.04.007

[103] Scapini G, Rasslan R, Cayuela NC, Goes MA, Koike MK, Utiyama EM, et al. Hypertonic saline and pentoxifylline enhance survival, reducing apoptosis and oxidative stress in a rat model of strangulated closed loop small bowel obstruction. Clinics (São Paulo, Brazil). [Internet]. 2019;**74**. Available from: https://www.pubmed.ncbi.nlm.nih.gov/31188910/

[104] Matsuo H, Hirose H, Mori Y, Takagi H, Iwata H, Yamada T, et al. Experimental studies to estimate the intestinal viability in a rat strangulated ileus model using a dielectric parameter. Digestive Diseases and Sciences. 2004;**49**(4):633-638

[105] Fevang J, Øvrebø K, Grong K, Svanes K. Fluid resuscitation improves intestinal blood flow and reduces the mucosal damage associated with strangulation obstruction in pigs. The Journal of Surgical Research. 2004;**117**(2):187-194

[106] Doğan G, İpek H, Baş Y, Doğan G, Kayır S. Experimental study on prophylactic effects of vardenafil in ischemia–reperfusion model with intestinal volvulus injury in rats. Journal of Pediatric Surgery. [Internet]. 2019;**54**(10):2172-2177. Available from: https://www.linkinghub.elsevier.com/retrieve/pii/S0022346819301241

[107] Darien BJ, Stone WC, Dubielzig RR, Clayton MK. Morphologic Changes of the Ascending Colon during Experimental Ischemia and Reperfusion in Ponies. Veterinary Pathology. [Internet]. May 1995;**32**(3):280-288. DOI: 101177/030098589503200310. PMID: 7604495

[108] Louw JH, Barnard CN. Congenital intestinal atresia observations on its origin. Lancet. 1955;**266**(6899):1065-1067

[109] Koga Y, Hayashida Y, Ikeda K, Inokuchi K, Hashimoto N. Intestinal atresia in fetal dogs produced by localized ligation of mesenteric vessels. Journal of Pediatric Surgery. 1975;**10**(6):949-953

[110] Patricolo M, Noia G, Rossi L, Zangari A, Pomini F, Catesini C, et al. An experimental animal model of intestinal obstruction to simulate in utero therapy for Jejunoileal Atresia. Fetal Diagnosis and Therapy. [Internet]. 1998;**13**(5):298-301. DOI: 10.1159/000020857

[111] Baglaj SM, Czernik J, Kuryszko J, Kuropka P. Natural history of experimental intestinal atresia: Morphologic and ultrastructural study. Journal of Pediatric Surgery. 2001;**36**(9):1428-1434

[112] Khen-Dunlop N, Fourcade L, Sauvat F, de Lambert G, Victor A, Cerf-Bensussan N, et al. Surgical experimental jejunoileal atresia in rat embryo. Journal of Pediatric Surgery. 2009;**44**(9):1725-1729

[113] Ballouhey Q, Fourcade L, Richard L, Bellet C, El Hamel C, Vallat JM, et al. Epithelial changes of congenital intestinal obstruction in a rat model. PLoS One. [Internet]. 2020;**15**(4). Available from: /pmc/articles/PMC7192479/

[114] Fourcade LM, Mousseau Y, Sauvat F, Khen-Dunlop N, Cerf-Bensussan N, Sarnacki S, et al. A new rat model of prenatal bowel obstruction: Development and early assessment. Journal of Pediatric Surgery. [Internet]. 2010;**45**(3):499-506. Available from: http://www.jpedsurg.org/article/S0022346809005910/fulltext

[115] Schoenberg RA, Kluth D. Experimental small bowel obstruction in chick embryos: Effects on the developing enteric nervous system. Journal of Pediatric Surgery. [Internet]. 2002;**37**(5):735-740. Available from: http://www.jpedsurg.org/article/S0022346802026441/fulltext

[116] Ordorica-Flores R, Orpinel-Armendariz E, Rodríguez-Reyna R, Pérez-Escamirosa F, Castro-Luna R, Minor-Martínez A, et al. Development and preliminary validation of a rabbit model of duodenal atresia for training in pediatric surgical skills. Surgical Innovation. 2019;**26**(6):738-743

[117] Merei JM. Notochord-gut failure of detachment and intestinal atresia. Pediatric Surgery International. 2004;**20**(6):439-443

[118] Mortell A, Montedonico S, Puri P. Animal models in pediatric surgery. Pediatric Surgery International. 2006;**22**(2):111-128

[119] Basson MD, Fielding LP, Bilchik AJ, Zucker KA, Ballantyne GH, Sussman J, et al. Does vasoactive intestinal polypeptide mediate the pathophysiology of bowel obstruction? American Journal of Surgery. [Internet]. 1989;**157**(1):109-115. Available from: https://www.pubmed.ncbi.nlm.nih.gov/2910115/

[120] Yang S, Shen L, Jin Y, Liu J, Gao J, Li D. Effect of Dachengqi decoction on NF-κB p65 expression in lung of rats with partial intestinal obstruction and the underlying mechanism. Journal of Huazhong University of Science and Technology. Medical Sciences. 2010;**30**(2):217-221

[121] Gálvez Y, Škába R, Vajtrová R, Frantlová A, Herget J. Evidence of secondary neuronal intestinal dysplasia in a rat model of chronic intestinal obstruction. Journal of Investigative Surgery. 2004;**17**(1):31-39

[122] Morel P, Alexander-Williams J, Rohner A. Relation between flow-pressure-diameter studies in experimental stenosis of rabbit and human small bowel. Gut. 1990;**31**(8):875-878

[123] Won KJ, Suzuki T, Hori M, Ozaki H. Motility disorder in experimentally obstructed intestine: Relationship between muscularis inflammation and disruption of the ICC network. Neurogastroenterology and Motility. 2006;**18**(1):53-61

[124] Sun D, Zhao J, Liao D, Chen P, Gregersen H. Shear modulus of the partially obstructed rat small intestine. Annals of Biomedical Engineering. [Internet]. 2017;**45**(4):1069-1082. Available from: https://www.link.springer.com/article/10.1007/s10439-016-1739-7

[125] Yuan ML, Yang Z, Li YC, Shi LL, Guo JL, Huang YQ, et al. Comparison of different methods of intestinal obstruction in a rat model. World J Gastroenterol [Internet]. 2013

[cited 2024 Mar 4];**19**(5):692-705. DOI: 10.3748/wjg.v19.i5.692

[126] Zhao J, Liao D, Yang J, Gregersen H. Phasic and tonic smooth muscle function of the partially obstructed guinea pig intestine. Journal of Biomedicine & Biotechnology. 2011;**2011**:1-9

[127] Ha SE, Wei L, Jorgensen BG, Lee MY, Park PJ, Poudrier SM, et al. A mouse model of intestinal partial obstruction. Journal of Visualized Experiments. 2018;**133**:1-7

[128] Wu B, Liu L, Gao H, Sun H, Xue H, Li X, et al. Distribution of interstitial cells of Cajal in Meriones unguiculatus and alterations in the development of incomplete intestinal obstruction. Histology and Histopathology. 2013;**28**(12):1567-1575

[129] Watanabe Y. Quantitative evaluation of experimental ischemic colitis correlated with the degree of artificial bowel obstruction in rats. Gastroenterologia Japonica. 1987;**22**(5):578-587

[130] Ekblad E, Sjuve R, Arner A, Sundler F. Enteric neuronal plasticity and a reduced number of interstitialcells of Cajal in hypertrophic rat ileum. Gut. [Internet]. 1998;**42**(6):836. Available from: /pmc/articles/PMC1727150/

[131] Bertoni S, Gabella G, Ghizzardi P, Ballabeni V, Impicciatore M, Lagrasta C, et al. Motor responses of rat hypertrophic intestine following chronic obstruction. Neurogastroenterology and Motility. 2004;**16**(3):365-374

[132] Bertoni S, Gabella G, Ballabeni V, Ghirardi A, Impicciatore M, Barocelli E. Plasticity of rat small intestine after removal of a chronic mechanical obstruction. Neurogastroenterol Motil [Internet]. 1 Sep 2006 [cited 2024 Mar 4];**18**(9):862-872. Available from: https://onlinelibrary.wiley.com/doi/full/10.1111/j.1365-2982.2006.00818.x

[133] Earlam RJ. Ganglion cell changes in experimental stenosis of the gut. Gut. 1971;**12**:393-398

[134] Wang Q, Zang J, Huang X, Lu H, Xu W, Chen J. Colonic dysmotility in murine partial colonic obstruction due to functional changes in interstitial cells. Journal of Neurogastroenterology and Motility. 2019;**25**(4):589-601

[135] Papanicolaou G, Nikas D, Ahn Y, Condos S, Fielding LP. Regional blood flow and water content of the obstructed small intestine. Archives of Surgery. [Internet]. 1985;**120**(8):926-932. Available from: https://www.jamanetwork.com/journals/jamasurgery/fullarticle/591177

[136] Chang IY, Glasgow NJ, Takayama I, Horiguchi K, Sanders KM, Ward SM. Loss of interstitial cells of Cajal and development of electrical dysfunction in murine small bowel obstruction. The Journal of Physiology. 2001;**536**(2):555-568

[137] Huang TY, Hanani M. Morphological and electrophysiological changes in mouse dorsal root ganglia after partial colonic obstruction. American Journal of Physiology. Gastrointestinal and Liver Physiology. 2005;**289**(4 52-4):670-678

[138] Collins J, Vicente Y, Georgeson K, Kelly D. Partial intestinal obstruction induces substantial mucosal proliferation in the pig. Journal of Pediatric Surgery. [Internet]. 1996;**31**(3):415-419. DOI: 10.1016/S0022-3468(96)90750-2%5Cn. Available from: http://www.sfx.library.uu.nl/utrecht?sid=EMBASE&issn=00223468&id=doi:10.1016%2FS0022-3468%2896%2990750-2&atitle=Partial+in

[139] Hiratsuka T, Inomata M. A novel animal model of colonic stenosis to aid the development of new stents for colon strictures. Surgical Endoscopy. [Internet]. 2022;**36**(5):3152-3159. DOI: 10.1007/s00464-021-08618-4

[140] Putranto AS, Suyatna FD, Soetikno V, Moenadjat Y. Novel and simple method using cable ties to induce intestinal strangulation in a rat model. Medical Journal of Indonesia. [Internet]. 2022;**31**(2):91-95. DOI: 10.13181/mji.oa.225799

[141] Lukas M, Kolar M, Ryska O, Juhas S, Juhasova J, Kalvach J, et al. Novel porcine model of Crohn's disease anastomotic stricture suitable for evaluation and training of advanced endoscopic techniques. Gastrointestinal Endoscopy. 2021;**93**(1):250-256

[142] Sun D, Zhao J, Liao D, Huang Z, Gregersen H. The turning point for morphomechanical remodeling during complete intestinal obstruction in rats occurs after 12-24 h. Annals of Biomedical Engineering. [Internet]. 2018;**46**(5):705-716. Available from: https://www.link.springer.com/article/10.1007/s10439-018-1992-z

[143] Storey BG, Lawrenson KB, Chant S, Stephens FO. Factors influencing the incidence of postoperative abdominal adhesions: An experimental study. The Australian and New Zealand Journal of Surgery. [Internet]. 1971;**40**(4):388-390. DOI: 10.1111/j.1445-2197.1971.tb04099.x

[144] Schade DS, Williamson JR. The pathogenesis of peritoneal adhesions: An ultrastructural study. Annals of Surgery. April 1968;**167**(4):500-510

[145] Moreno Egea A, Aguayo Albasini JL, Zambudio Carmona G, Parrilla PP. Adhesion response to different forms of treating a peritoneal lesion: An experimental study in rats. Digestive Surgery. [Internet]. 1995;**12**(6):334-337. Available from: https://www.karger.com/Article/FullText/172385

[146] McEntee GP, Stuart RC, Byrne PJ, Leen E, Hennessy TP. Experimental study of starch-induced intraperitoneal adhesions. The British Journal of Surgery. [Internet]. 2005;**77**(10):1113-1114. DOI: 10.1002/bjs.1800771012

[147] Barambio J, García-Arranz M, Campos PV, Pinto JFV, Clemente LV, Gómez-Heras SG, et al. Chemical scalpel: An experimental collagenase-based treatment for peritoneal adhesions. Biology (Basel). [Internet]. 2022;**11**(8). Available from: https://www.pubmed.ncbi.nlm.nih.gov/36009786/

[148] Marshall CD, Hu MS, Leavitt T, Barnes LA, Cheung ATM, Malhotra S, et al. Creation of abdominal adhesions in mice. Journal of Visualized Experiments : JoVE. [Internet]. 2016;**2016**(114):e54450. Available from: https://www.jove.com/v/54450/creation-of-abdominal-adhesions-in-mice

[149] Vediappan RS, Bennett C, Bassiouni A, Smith M, Finnie J, Trochsler M, et al. A novel rat model to test intra-abdominal anti-adhesive therapy. Frontiers in Surgery. [Internet]. 2020;**7**:12. Available from: /pmc/articles/PMC7158702/

[150] Kalff JC, Schraut WH, Simmons RL, Bauer AJ. Surgical manipulation of the gut elicits an intestinal muscularis inflammatory response resulting in postsurgical ileus. Annals of Surgery. Nov 1998;**228**(5):652-663. DOI: 10.1097/00000658-199811000-00004. PMID: 9833803

[151] Strik C, Wever KE, Stommel MWJ, van Goor H, ten Broek RPG. Adhesion reformation and the limited translational

value of experiments with adhesion barriers: A systematic review and meta-analysis of animal models. Scientific Reports. [Internet]. 2019;**9**(1):1-10. DOI: 10.1038/s41598-019-52457-2

[152] Gaertner WB, Hagerman GF, Felemovicius I, Bonsack ME, Delaney JP. Two experimental models for generating abdominal adhesions. The Journal of Surgical Research. 2008;**146**(2):241-245

[153] Schippers E, Tittel A, Öttinger A, Schumpelick V. Laparoscopy versus laparotomy: Comparison of adhesion-formation after bowel resection in a Canine model. Digestive Surgery. [Internet]. 1998;**15**(2):145-147. DOI: 10.1159/000018608

[154] Iijima N, Inoue K, Yamamoto T, Gomi F. Experimental studies on intra-abdominal adhesions. Postgraduate Medical Journal. [Internet]. 1970;**46**(535): 278-282. Available from: https://www.pubmed.ncbi.nlm.nih.gov/5448376/

[155] Ellis H. The aetiology of post-operative abdominal adhesions an experimental study. The British Journal of Surgery. [Internet]. 2005;**50**(219):10-16. DOI: 10.1002/bjs.18005021904

[156] McBride WH, Mason KA, Davis C, Withers HR, Smathers JB. Adhesion formation in experimental chronic radiation enteropathy. International Journal of Radiation Oncology, Biology, Physics. [Internet]. 1989;**16**(3):737-743. Available from: http://www.redjournal.org/article/0360301689904938/fulltext

[157] Seitz HM, Schenker JG, Epstein S, Garcia CR. Postoperative intraperitoneal adhesions: A double-blind assessment of their prevention in the monkey. Fertility and Sterility. 1973;**24**(12):935-940

[158] Evrard VAC, De Bellis A, Boeckx W, Brosens IA. Surgery: Peritoneal healing after fibrin glue application: A comparative study in a rat model. Human Reproduction. [Internet]. 1996;**11**(9):1877-1880. DOI: 10.1093/oxfordjournals.humrep.a019510

[159] Beach EC, De JO. Ileus. In: Encycl Gastroenterol. 2nd ed. [Internet]. StatPearls Publishing LLC. 2024. pp. 241-243. Available from: https://www.ncbi.nlm.nih.gov/books/NBK558937/

[160] Woodfork K. Paralytic ileus. xPharm: The Comprehensive Pharmacology Reference. 2007;**2004**:1-9

[161] Schwarz NT, Beer-Stolz D, Simmons RL, Bauer AJ. Pathogenesis of paralytic ileus: Intestinal manipulation opens a transient pathway between the intestinal lumen and the leukocytic infiltrate of the jejunal muscularis. Annals of Surgery. [Internet]. 2002;**235**(1):31. Available from: /pmc/articles/PMC1422393/

[162] Yu L, Y, Hua CM, Goetz B, Qiu CC, Jing FY, Chen CJ, et al. The dual effect of cannabinoid receptor-1 deficiency on the murine postoperative ileus. PLoS One. [Internet]. 2013;**8**(7). Available from: https://www.pubmed.ncbi.nlm.nih.gov/23844009/

[163] Morris IR, Darby CF, Hammond P, Taylor I. Changes in small bowel myoelectrical activity following laparotomy. The British Journal of Surgery. [Internet]. 2005;**70**(9):547-548. DOI: 10.1002/bjs.1800700913

[164] Vilz TO, Overhaus M, Stoffels B, von Websky M, Kalff JC, Wehner S. Functional assessment of intestinal motility and gut wall inflammation in rodents: Analyses in a standardized model of intestinal manipulation. Journal of Visualized Experiments. 2012;**67**:3-9

[165] Hartmann L, Arndt M, Hahn EM, Mueller MH, Kreis ME, Hering NA. Effect of bowel preparation on intestinal permeability and inflammatory response during postoperative ileus in mice. Surgery (United States). [Internet]. 2021;**170**(5):1442-1447. Available from: http://www.surgjournal.com/article/S0039606021004335/fulltext

[166] The FO, Cailotto C, Van Der Vliet J, De Jonge WJ, Bennink RJ, Buijs RM, et al. Central activation of the cholinergic anti-inflammatory pathway reduces surgical inflammation in experimental post-operative ileus. British Journal of Pharmacology. [Internet]. 2011;**163**(5):1007-1016. DOI: 10.1111/j.1476-5381.2011.01296.x

[167] De Winter BY, Boeckxstaens GE, De Man JG, Moreels TG, Herman AG, Pelckmans PA. Effect of adrenergic and nitrergic blockade on experimental ileus in rats. British Journal of Pharmacology. 1997;**120**(3):464-468

[168] De Winter BY, Boeckxstaens GE, De Man JG, Moreels TG, Herman AG, Pelckmans PA. Differential effect of indomethacin and ketorolac on postoperative ileus in rats. European Journal of Pharmacology. 1998;**344**(1):71-76

[169] Matsumoto K, Kawanaka H, Hori M, Kusamori K, Utsumi D, Tsukahara T, et al. Role of transient receptor potential melastatin 2 in surgical inflammation and dysmotility in a mouse model of postoperative ileus. American Journal of Physiology. Gastrointestinal and Liver Physiology. 2018;**315**(1):G104-G116

[170] Sun Y, Shi H, Hong Z, Chi P. Inhibition of JAK1 mitigates postoperative ileus in mice. Surgery (United States). [Internet]. 2019;**166**(6):1048-1054. DOI: 10.1016/j.surg.2019.07.016

[171] Gerring EE, Hunt JM. Pathophysiology of equine postoperative ileus: Effect of adrenergic blockade, parasympathetic stimulation and metoclopramide in an experimental model. Equine Veterinary Journal. 1986;**18**(4):249-255

[172] van Bree SHW, Nemethova A, van de Bovenkamp FS, Gomez-Pinilla P, Elbers L, Di Giovangiulio M, et al. Novel method for studying postoperative ileus in mice. International Journal of Physiology, Pathophysiology and Pharmacology. [Internet]. 2012;**4**(4):219. Available from: /pmc/articles/PMC3544220/

[173] Ninomiya N, Nemoto K, Okamura T, Suzuki H, Yamamoto Y. A novel experimental method for the study of intestinal paralysis due to endotoxicosis. Nihon Kyukyu Igakukai Zasshi. 2003;**14**(5):241-250

[174] Holzer P, Lippe IT, Holzer-Petsche U. Inhibition of gastrointestinal transit due to surgical trauma or peritoneal irritation is reduced in capsaicin-treated rats. Gastroenterology. 1986;**91**(2):360-363

[175] Friese N, Chevalier E, Angel F, Pascaud X, Junien JL, Dahl SG, et al. Reversal by K-agonists of peritoneal irritation-induced ileus and visceral pain in rats. Life Sciences. 1997;**60**(9):625-634

[176] Kobayashi I, Kajisa M, Farid AS, Yamanaka A, Horii Y. Paralytic ileus and subsequent death caused by enteric parasite, strongyloides papillosus, in mongolian gerbils. Veterinary Parasitology. [Internet]. 2009;**162**(1-2):100-105. Available from: https://www.pubmed.ncbi.nlm.nih.gov/19303216/

[177] Vera G, López-Pérez AE, Uranga JA, Girón R, Martín-Fontelles MI, Abalo R.

Involvement of cannabinoid signaling in vincristine-induced gastrointestinal dysmotility in the rat. Frontiers in Pharmacology. 2017;**8**(Feb):242099. Available from: https://www.frontiersin.org

[178] Chang H, Li S, Li Y, Hu H, Cheng B, Miao J, et al. Effect of sedation with dexmedetomidine or propofol on gastrointestinal motility in lipopolysaccharide-induced endotoxemic mice. BMC Anesthesiology. [Internet]. 2020;**20**(1). Available from: https://www.pubmed.ncbi.nlm.nih.gov/32894042/

[179] De Filippis D, Iuvone T, D'Amico A, Esposito G, Steardo L, Herman AG, et al. Effect of cannabidiol on sepsis-induced motility disturbances in mice: Involvement of CB1 receptors and fatty acid amide hydrolase. Neurogastroenterology and Motility. 2008;**20**(8):919-927

[180] Cai L, Rodgers E, Schoenmann N, Raju RP. advances in rodent experimental models of sepsis. International Journal of Molecular Sciences. 2023;**24**:9578. [Internet]. Available from: https://www.mdpi.com/1422-0067/24/11/9578/htm

[181] Udassin R, Eimerl D, Schiffman J, Haskel Y. Epidural anesthesia accelerates the recovery of postischemic bowel motility in the rat. Anesthesiology. [Internet]. 1994;**80**(4):832-836. DOI:10.1097/00000542-199404000-00016

[182] Dejager L, Pinheiro I, Dejonckheere E, Libert C. Cecal ligation and puncture: The gold standard model for polymicrobial sepsis? Trends in Microbiology. [Internet]. 2011;**19**(4):198-208. Available from: http://www.cell.com/article/S0966842X11000035/fulltext

[183] Overhaus M, Tögel S, Pezzone MA, Bauer AJ. Mechanisms of polymicrobial sepsis-induced ileus. American Journal of Physiology. Gastrointestinal and Liver Physiology. [Internet]. 2004;**287**(3):50-53. Available from: https://www.journals.physiology.org/doi/10.1152/ajpgi.00359.2003

[184] Kylolu G, Kaya T, Bagcivan I, Yildiz T. Effect of L-NAME on decreased ileal muscle contractility induced by peritonitis in rats. Journal of Pediatric Surgery. [Internet]. 2002;**37**(6):901-905. Available from: http://www.jpedsurg.org/article/S0022346802778895/fulltext

[185] Teke Z, Sacar M, Yenisey C, Atalay AO, Kavak T, Erdem E. Activated protein C attenuates intestinal mucosal injury after mesenteric ischemia/reperfusion. The Journal of Surgical Research. 2008;**149**(2):219-230

Chapter 3

# Exploring the Prospect and Significance of Nematodes (*Caenorhabditis elegans*) as an Animal Model for Translational Neurobiology Research

*Gideon Olamilekan Oluwatunase, Patrick Godwin Okwule, Olayinka Olawale Asafa, David Anuoluwapo Oyeniran and Onome Bright Ogenetega*

## Abstract

The model organism *Caenorhabditis elegans* (*C. elegans*) is a microscopic roundworm discovered in 1900, which is widely becoming an acceptable animal model of human diseases in the scientific community, particularly in neurobiology and genetic research. *C. elegans* is considered of great interest to neuroscientists for studying basic biological processes and understanding neurodegenerative and neurodevelopmental disorders because of its small size, fully sequenced genome, simple and transparent body structure, and short life span. This chapter highlights the use of *C. elegans* in modeling diseases such as Alzheimer's, Parkinson's, and amyotrophic lateral sclerosis (ALS), alongside its contributions to neurodevelopmental and behavioral studies. Furthermore, advanced tools like CRISPR/Cas9 and RNA interference have augmented its utility, making it invaluable for therapeutic discovery and understanding gene-environment interactions. Challenges and future prospects are also presented hereon.

**Keywords:** *Caenorhabditis elegans*, neurobiology, translational research, neurodegenerative diseases, animal model

## 1. Introduction

### 1.1 Translational neurobiology

Translational Neurobiology is a field of study that employs neuroscientific research to develop or translate into novel therapeutics and clinical applications for

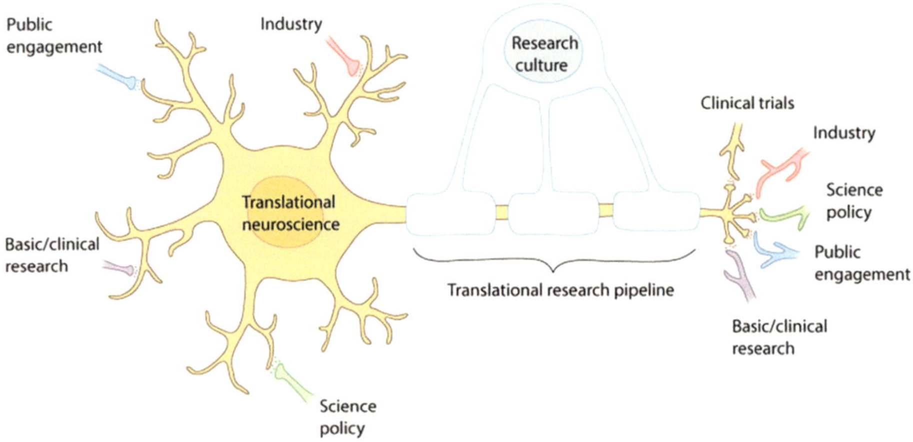


**Figure 1.**
*Image showing the scope of neurobiology and neuroscientific studies [1].*

diseases of the nervous system. Translational neurobiology is a methodical, theory-driven approach that holds promise for advancing and utilizing fundamental and applied neuroscientific knowledge to support the creation and improvement of clinical and public health interventions.

By translating basic science knowledge into interventions and applications for treating human disease, translational neuroscience seeks to address neurological or mental health issues [1].

The goal of the multidisciplinary discipline of translational neurobiology is to apply basic neuroscience research findings to the treatment, prevention, or diagnosis of neurological and psychiatric illnesses. The objective is to "translate" findings from laboratory research to patient bedside treatment. This area of study fills the knowledge gap between basic neuroscience, which studies the nervous system's operation, and clinical practice, which treats illnesses of the brain. At the moment, many brain illnesses are incurable. It has been proposed that this could be altered by approaching neuroscientific research with a "translational" mindset (**Figure 1**) [1].

## 2. Animal model of human disease

For many years, animal models have been used extensively in biomedical research because they provide critical and valuable insights into the pathogenesis of human diseases and the development of novel therapeutic interventions. Choosing animal models that have similarities to human tissues or disease conditions is crucial in biomedical research. Some models have the benefit of having more rapid mitotic, developmental, aging, or reproductive cycles than humans, which enhances biomedical research.

Animal models serve as an essential bridge between basic scientific discoveries and clinical applications. Many features of human biology are shared across species, and animals can mimic human diseases with respect to symptoms and underlying biological mechanisms. In this chapter, we will discuss *Caenorhabditis elegans* extensively as one of the emerging animal models used in translational neurobiological research.

## 2.1 *Caenorhabditis elegans*

The free-living nematode *Caenorhabditis elegans*, sometimes known as *C. elegans*, is widely used as an animal model in studies related to aging and neurodegenerative illnesses, such as Parkinson's and Alzheimer's, among others. This tiny nematode has several advantages over typical animal models, including a small body size, a short lifetime, a fully sequenced genome, and over 65% of the genes linked to human disease. This organism is a perfect living system for studying neurobiology because of all these qualities (**Figure 2**) [3].

## 2.2 General characteristics of *C. elegans*

- *Life cycle*

  *C. elegans* has three disc larval stages, and adulthood. In a laboratory context, *C. elegans* can be kept at 16 to 25°C, although 20°C is the usual temperature. When environmental circumstances are unfavorable, the larval worms release a pheromone that causes them to transition into a dauer stage, where they can withstand harsh conditions for several months. *C. elegans* has a mouth, throat, gut, gonad, and cuticle as part of its fundamental anatomy [4].

- *Reproductive system*

  For C. elegans, there are two sexes: hermaphrodite and male. The reproductive system of the self-fertilizing hermaphrodite consists of somatic gonad, the germ line and the egg-laying apparatus. The germ line produces both sperm and oocytes, respectively.

- *Nervous system*

  For C. elegans, an adult hermaphrodite has 302 neurons that belong to two distinct and independent nervous systems: a large somatic nervous system (282 neurons) and a small pharyngeal nervous system (20 neurons). The nervous system of C. elegans is the most complex organ with the highest cellular variety (at least 118 distinct neuron

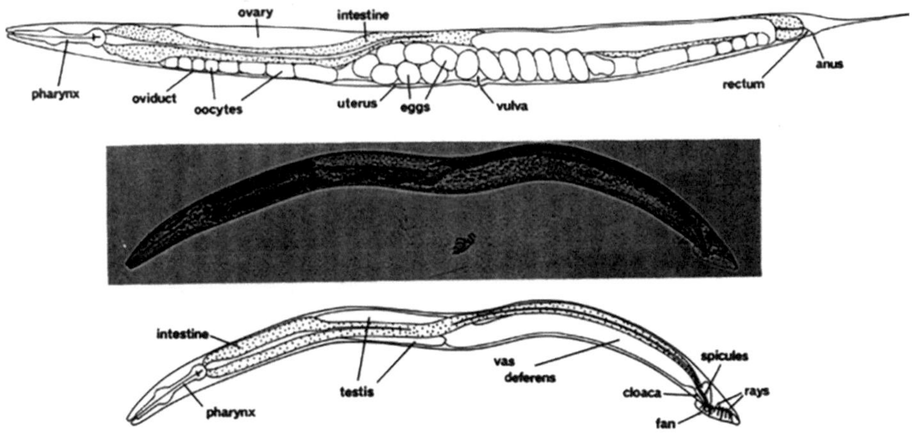


**Figure 2.**
*Image showing the internal anatomy of* C. elegans *including the reproductive system [2].*

classes) and the greatest number of cells (it makes up 37% of a hermaphrodite's somatic cells). Furthermore, C. elegans contains nearly every gene family implicated in mammalian neuron function [4]. Consequently, a variety of neurological illnesses, such as Parkinson's disease, Alzheimer's disease, and other neurodegenerative disorders, have been studied using C. elegans as an experimental model [5].

## 3. Advantages of *C. elegans* as a model organism

In the second half of the twentieth century, Sydney Brenner introduced the soil nematode *Caenorhabditis elegans* as a model organism for studying development and neurobiology. This pioneering work laid the groundwork for its extensive use in various biological research areas today, including apoptosis, cell signaling, cell cycle regulation, cell polarity, gene expression, metabolism, aging, and sex determination [6]. The simplicity of *C. elegans*' anatomy and genetics, combined with its well-mapped neural circuitry, makes it an invaluable tool for uncovering fundamental biological processes.

The genetic tractability of *C. elegans* has enabled precise manipulation of genes and facilitated large-scale genetic screens. The ability to knock out genes or induce loss-of-function mutations has significantly advanced our understanding of the molecular pathways involved in various biological diseases [7]. Given the complexity of the human genome, which poses challenges for studying disease pathogenesis, *C. elegans* offers a simpler alternative. Its complete genome has been sequenced and found to share substantial homology with the human genome, making it a focal point for experimental research [8]. Many essential biological and stress responses in higher organisms, including humans, are highly conserved in *C. elegans*. Depending on the bioinformatics techniques used, researchers have discovered homologs for around 60–80% of human genes in *C. elegans* [9]. Additionally, 12 of the 17 identified signal transduction pathways are also shared between *C. elegans* and humans.

Recent advances in genetic technologies, such as RNA interference (RNAi) and CRISPR/Cas9 genome editing, have further expanded opportunities for gene manipulation in *C. elegans*, enhancing our ability to investigate human biology [10]. Many diseases arise from gene mutations during embryogenesis, which may be influenced by environmental factors. Understanding the interplay between gene mutations and environmental influences is crucial for elucidating disease pathogenesis and progression. *C. elegans* is effectively utilized to model these interactions in various neurodevelopmental and neurodegenerative diseases, providing insights into how genetic and environmental factors converge to affect health outcomes [11].

The short life cycle of the worm is another great advantage to medicine and biological science. Studying age and its related diseases has not been without a challenge. The introduction of *C. elegans* has led to its use as a good model to understand the aging process. This is possible because the worm has a life cycle of approximately 2.5–4 days and consists of several distinct stages: embryogenesis (approximately 16 hours), four larval stages (L1 to L4, totaling around 28 hours), and the adult molt, which lasts about 12 hours [12–14]. During the late L1 larval stage, adverse environmental conditions—such as starvation, high population density, or elevated temperatures—can trigger a developmental arrest phase known as dauer. Dauer larvae possess remarkable resilience, allowing them to survive in these stressful conditions for several months. When they eventually encounter more favorable environments, they can resume their reproductive development and continue their life cycle [13].

The short lifespan and small size of *Caenorhabditis elegans* make it an ideal model organism for testing anti-aging drugs. Its low maintenance cost, combined with the ability to screen the entire organism for experimental outcomes, enhances its utility in research. The worm's transparent body facilitates direct observation of internal organs, while its high fertility rate allows for the rapid generation of offspring [15]. Additionally, the availability of advanced molecular biology tools for gene manipulation further establishes this nematode as a valuable model for studying aging-related mutations, as it shares many biological responses with humans. Furthermore, experiments involving *C. elegans* do not require ethical approval for now, providing a significant advantage over other experimental models [16]. This unique characteristic enables researchers to conduct studies more efficiently and with fewer regulations, making *C. elegans* a preferred choice for aging research and other biological investigations [16].

*C. elegans* offers opportunities to directly test the effects of drugs to asses and remove compounds that are toxic to development. This nematode is also invaluable for genetic analysis and exploring chemical interventions aimed at enhancing longevity [17]. Currently, various assays are being developed for high-throughput screening of anti-aging compounds. Research has identified several compounds that significantly extend the lifespan of *C. elegans*, particularly those linked to mutations in the age-1 (PI3K) and daf-2 (INSR) genes, as well as reductions in the daf-16 (FOXO) gene [18]. For instance, Kumar *et al.* (2015) Kumar's study found that treatment with 25 and 50 μM silymarin increased the mean lifespan of *C. elegans* by 10.1 and 24.8%, respectively, compared to untreated controls [19]. Additionally, another study demonstrated that fullerenol reduced endogenous levels of reactive oxygen species (ROS) and protected *C. elegans* by upregulating stress-related genes under stressful conditions in a DAF-16-dependent manner, thereby enhancing lifespan [20].

The field of neurobiology has been improved since the introduction of the *C. elegans*. The nervous system of this worm is known to contain only 302 neurons. The interesting advantage of this is that all the neurons have been fully mapped out, forming a complete neuron connectome [21]. This simple neural wiring makes it a tool to understand human neural connections. The neural function and signaling pathway found in this nematode are similar to that of the mechanism found in other higher mammals including humans [6]. Moreso, human neuronal activities cannot be observed from the outside but from the transparent body. *C. elegans* gives a real-time evaluation of what is going on in the nervous system, which can give insight into the neural activity of the human brain [22]. This worm has also been used to understand human behavior and factors contributing to changing behavior [23].

## 4. Applications of *Caenorhabditis elegans* in neurobiology research

### 4.1 Modeling neurodegenerative diseases

Research in neurodegenerative diseases such as Alzheimer's, Parkinson's, and Huntington's diseases benefits significantly from the use of model organisms like *C. elegans*. These diseases involve the progressive deterioration of neuronal function, leading to cognitive and motor impairments. By introducing genetic mutations in *C. elegans* that mimic those seen in human patients, researchers have developed models that replicate key aspects of these conditions. For example, *C. elegans* models of Alzheimer's disease typically involve the expression of amyloid-beta, which leads to neurodegeneration in a manner similar to that seen in human patients [24]. *C. elegans* is also a powerful model for studying Parkinson's disease through the manipulation of dopaminergic neurons and Huntington's disease by introducing polyglutamine expansions, allowing researchers to investigate mechanisms of disease progression and potential therapeutic interventions [25]. The use of *C. elegans* provides critical insights into neurodegenerative diseases because of the organism's well-characterized nervous system, ease of genetic manipulation, and short lifespan. These models have revealed important information about the genetic and molecular pathways involved in protein aggregation, neuronal death, and the role of cellular stress responses [26].

Huntington's disease (HD) is a hereditary neurodegenerative condition that is inherited in an autosomal dominant pattern. It is caused by an expansion of CAG repeats in the huntingtin (HTT) gene, resulting in the production of a mutant HTT protein with elongated polyglutamine sequences. The abnormal accumulation of this protein can lead to neuronal death, contributing to the gradual deterioration of motor control and cognitive functions associated with this disease [27, 28]. Although *C. elegans* lacks an HTT ortholog, various HD models have been developed that express human polyQ proteins. This has allowed researchers to study the accumulation of these proteins and their associated toxic effects during aging [29, 30]. Additionally, mechanisms such as impaired vesicle trafficking, heat shock response, and mitochondrial stress have been linked to α-syn toxicity in *C. elegans* [31]. The CA150 or CRE 60 Huntingtin-interacting proteins have been identified in *C. elegans* during yeast-two hybrid screening of the worm proteome, which is identical to the human Huntingtin homologs [30]. This model of the worm has advanced the study of understanding the pathogenesis of Huntington's disease as it provides an opportunity for a whole animal study suggesting new therapeutic interventions for drug screening [9]. Studies using *C. elegans* have uncovered potential factors that may influence polyQ aggregation and its associated toxicity. Different substances have been used on the *C. elegans* model of Huntington such as diphenyl diselenide and Rutin have been found to show cytoprotective effects through enhanced autophagy and antioxidant activity [32, 33].

### 4.2 Neurodevelopmental studies

Neuronal development and differentiation are fundamental processes in neurobiology, and *C. elegans* serves as an excellent model for studying these processes. The transparency of *C. elegans* enables researchers to track neuronal development from the embryo to the adult stage. Studies in *C. elegans* have contributed to understanding how neurons differentiate, form synapses, and establish functional neural circuits [34]. Synapse formation and function, in particular, have been extensively studied using *C. elegans*, leading to significant discoveries about the roles of synaptic proteins and neurotransmitter release mechanisms. Additionally, research using *C. elegans* has helped elucidate the genetic basis of various neurodevelopmental disorders, including autism spectrum disorder and intellectual disabilities [35].

### 4.3 Behavioral studies

*C. elegans* has also been widely used in behavioral studies, particularly in understanding locomotion, learning, memory, and sensory-motor functions. The worm's simple nervous system, composed of 302 neurons, allows for detailed mapping of neural circuits involved in behavior. Locomotion studies in *C. elegans* have provided insights into the functioning of motor neurons and muscle activity [36]. Researchers also study learning and memory using associative learning paradigms, where worms learn to associate specific stimuli, such as temperature or odors, with certain outcomes [37]. These studies help in understanding the neural basis of behavior, as well as the genetic and molecular mechanisms that underlie cognitive functions.

In terms of sensory and motor functions, *C. elegans* responds to various environmental cues, such as chemical signals and mechanical stimuli. The nervous system's ability to process these stimuli and generate appropriate behavioral responses is a key area of study, particularly for understanding sensory processing disorders and motor deficits [38].

## 5. Drug screening and toxicology

One of the most practical applications of *C. elegans* in neurobiology research is its use in high-throughput drug screening and toxicology studies because of its small size, short lifecycle, and ease of genetic manipulation. *C. elegans* is a valuable model for testing the neurotoxicity of compounds and assessing therapeutic efficacy in a living organism. This is particularly useful for screening large libraries of small molecules in neurodegenerative disease models [9]. The ability to test neuroprotective compounds in a live model allows researchers to identify potential drug candidates more quickly and assess their effects on neural function. Furthermore, *C. elegans* is used to study the effects of environmental toxins on the nervous system, offering insights into neurotoxicity and neural resilience [39].

## 6. Methodological approaches

### 6.1 Genetic and genomic tools

*Caenorhabditis elegans* (*C. elegans*) has become a powerful model organism due to its genetic tractability. Techniques such as CRISPR-Cas9 and RNA interference (RNAi) are widely used for gene editing and silencing. CRISPR-Cas9 allows precise genome editing by creating double-strand breaks at specific locations, which are then repaired by the cell's machinery, leading to targeted mutations [40]. RNAi, on the other hand, involves the introduction of double-stranded RNA to silence specific genes, providing a robust method for studying gene function [41].

### 6.2 CRISPR-Cas9 technology

CRISPR-Cas9 has revolutionized the field of genetics, and its application in *C. elegans* research has been transformative. This technology allows for highly specific, targeted genome modifications by inducing double-strand breaks at precise loci. These breaks are repaired by the cell's endogenous repair mechanisms, either through homology-directed repair (HDR) or non-homologous end joining (NHEJ). CRISPR-Cas9 is particularly useful in creating mutant strains of *C. elegans* that can model human diseases, including neurodegenerative disorders. Mutants with altered genes, such as those involved in amyloid-beta production or alpha-synuclein aggregation, can be used to study the underlying mechanisms of Alzheimer's and Parkinson's diseases, respectively [40].

## 6.3 RNA interference (RNAi)

RNA interference (RNAi) is another powerful genetic tool in *C. elegans* research. This technique allows researchers to silence specific genes by introducing double-stranded RNA (dsRNA) molecules into the worms, triggering the degradation of the corresponding mRNA. RNAi has been instrumental in functional genomics, enabling the study of gene knockdown effects *in vivo* [42]. One of the key advantages of RNAi is its reversibility, which allows for temporal control of gene silencing. This is particularly important for studying genes that are essential for development, as the knockdown can be induced at specific stages to avoid lethal phenotypes.

## 6.4 Transgenesis and gene knockouts

Transgenic techniques, including the creation of gene knockout strains, are widely used in *C. elegans* to investigate the function of specific genes in neural development and disease. Gene knockouts can be achieved through targeted mutagenesis or CRISPR-Cas9-mediated gene editing, allowing researchers to study the effects of complete gene loss. Conversely, transgenic strains can be engineered to overexpress genes of interest, such as those encoding for neurotoxins or human disease-related proteins. These techniques have been invaluable in modeling neurological diseases and screening for potential therapeutic compounds.

## 6.5 Transcriptomics and proteomics

Transcriptomic and proteomic analyses in *C. elegans* provide comprehensive insights into gene expression and protein function. High-throughput sequencing technologies, such as RNA-Seq, allow for the quantification of gene expression levels across different conditions and developmental stages [39]. Proteomics, involving mass spectrometry, enables the identification and quantification of proteins, offering a deeper understanding of cellular processes and protein interactions [43].

## 6.6 Transcriptomics in *C. elegans*

Advances in high-throughput sequencing technologies, such as RNA sequencing (RNA-Seq), have made it possible to quantify gene expression across different tissues and life stages of *C. elegans*. This has been particularly useful in understanding how transcriptional networks are regulated in neurons and how they change in response to environmental stimuli or genetic mutations [39]. Transcriptomic analyses have shed light on the molecular changes that occur during neurodegeneration, providing potential biomarkers for early diagnosis and targets for intervention.

## 6.7 Proteomics and protein dynamics

Proteomics, the large-scale study of proteins, is another essential tool in *C. elegans* neurobiology research. Mass spectrometry-based proteomics allows for the identification and quantification of proteins in different tissues, providing insights into protein expression levels, post-translational modifications, and protein-protein interactions [43]. Proteomic studies in *C. elegans* have identified key proteins involved in synaptic transmission, neuronal development, and neurodegeneration, furthering our understanding of the molecular basis of neurological diseases.

1. *Imaging and electrophysiology*

   This combines with 3D cardiac computed tomography (CT), MRI, and other imaging modalities to provide accurate anatomical and electrical maps of the heart.

2. *Advanced microscopy techniques for live imaging*

   Live imaging techniques are indispensable for visualizing cellular and subcellular structures in *C. elegans*. Confocal and two-photon microscopy, in particular, allow for high-resolution, real-time imaging of neurons and other tissues in living worms. Researchers often use genetically encoded fluorescent proteins, such as GFP (green fluorescent protein), to label specific neurons or signaling molecules [44]. This enables the study of dynamic processes such as synaptic plasticity, neuronal migration, and axonal guidance. In addition to fluorescence-based imaging, calcium-sensitive indicators like GCaMP have been developed to monitor neural activity in response to external stimuli. Calcium imaging provides real-time data on neuronal activity, allowing researchers to map functional connectivity in the nervous system of *C. elegans*. These imaging techniques have been instrumental in uncovering how neural circuits are organized and how they adapt to changes in the environment or genetic perturbations [45].

3. *Electrophysiological methods for neural activity recording*

   Electrophysiology is a critical technique for studying the electrical properties of neurons in *C. elegans*. Electrophysiological techniques, including patch-clamp recordings and in vivo calcium imaging, are employed to measure neural activity in *C. elegans*. These methods enable the study of ion channel function, synaptic transmission, and neural circuit dynamics [46]. Whole-cell patch-clamp electrophysiology, in particular, provides high-resolution data on the electrical properties of neurons [47]. While the worm's small size poses challenges for traditional patch-clamp recordings, researchers have developed alternative methods, such as optogenetics and microelectrode arrays, to study neural activity. Optogenetics involves the use of light-sensitive proteins, such as channelrhodopsins, to control neuronal activity with precise spatial and temporal resolution. By using light to stimulate or inhibit specific neurons, researchers can investigate how individual neurons contribute to behavior and how neural circuits are wired. Electrophysiological studies in *C. elegans* have provided insights into the mechanisms of synaptic transmission, ion channel function, and neuronal excitability. These studies have been particularly useful for understanding how mutations in ion channels and synaptic proteins contribute to neurological diseases such as epilepsy, ALS, and Parkinson's disease (PD) [48].

4. *Standardized assays for measuring behavior and neural function*

   Behavioral assays in *C. elegans* are standardized to measure various aspects of neural function and behavior. Chemotaxis assays, for example, assess the nematode's ability to navigate chemical gradients, providing insights into sensory processing and neural circuitry [49]. Other assays, such as the 1-Nonanol and Aldicarb assays, are used to study neurotoxic effects and synaptic function [50].

5. *Computational tools for data analysis and interpretation*

Computational tools are integral to the analysis and interpretation of behavioral data in *C. elegans* research. Software platforms like WormLab and MATLAB are used to track and quantify nematode movements, enabling detailed analysis of behavioral patterns and neural responses. These tools facilitate the integration of large datasets, enhancing the reproducibility and accuracy of experimental findings.

## 7. Case studies

*Case Study 1:* Alzheimer's disease (AD) is characterized by insoluble Aβ (amyloid β-peptide) plaques and tau-associated neurofibrillary tangles. This occurs with the Aβ accumulation resulting from the enzymatic processing of the APP (amyloid precursor protein). Although *C. elegans* lacks clear orthologs for β-secretase and APP, however, it has been utilized to model AD through the transgenic expression of human amyloid-beta (Aβ) and tau proteins [51]. The modeling of AD in *C. elegans* is mainly based on the amyloid cascade hypothesis, with studies focusing on the toxic Aβ1-42 peptide, which has been found to induce neurodegeneration in specific neurons. This has made the quantitative analysis of modulators and genetic factors that influence Aβ toxicity. Possible [52, 53]. Recent research has applied the use of this worm in neuroprotection. Using this *C. elegans*, it was found that the protective genetic modifiers, such as PICALM [54], and other substances that degrade Aβ are potential therapeutic agents against AD [55, 56]. Transgenic models expressing human apolipoprotein E (APOE) alleles have shown that APOEε2 protects against amyloid β-peptide neurotoxicity while APOEε4 does not, thus recapitulating the clinical profile of APOE polymorphisms and providing an understanding of the mechanisms underlying AD [57, 58].

*Case Study 2:* Parkinson's disease is ranked the second most prevalent neurodegenerative disorder, affecting approximately 2% of individuals about the age of 65 years [59, 60]. Studies have shown that PD is linked to environmental factors, epigenetics, and genetics. This could be exposure to toxicants like pesticides, metals, and neurotoxins [59, 61]. Research using *C. elegans* has shown that exposure to rotenone, which is a mitochondrial complex I inhibitor, leads to dopaminergic neuron loss and other defects. Mitochonic acid 5 (MA-5) has demonstrated protective effects against mitochondrial damage and degeneration of these neurons [62]. PD symptoms are caused primarily by the loss of dopamine-producing neurons in the substantia nigra, a process marked by the presence of Lewy bodies containing alpha-synuclein [63]. However, many underlying molecular pathways remain unclear and more studies to understand the epigenetic mechanisms and potential therapeutic targets using the *C. elegans* are important. The α-synuclein (α-syn) protein, encoded by the PARK1/SNCA gene in humans, is predominantly located at presynaptic terminals and within the nucleus of neurons [64, 65]. *C. elegans* does not have a direct ortholog for this protein. Instead, nematode models of α-synuclein are created by expressing normal or disease-related variants of human α-synuclein in neuronal or non-neuronal cells. It is worth noting that the pathological characteristics of these models are primarily determined by the expression pattern of α-syn and not the specific type of protein (normal or disease-associated) [66].

*Case Study 3:* This study discusses translational insights into human neurological disorders. *C. elegans* has proven to be an invaluable model for studying the molecular mechanisms underlying human neurological disorders. In particular, it has been used to model Alzheimer's disease (AD), Parkinson's disease (PD), Huntington's disease (HD), and amyotrophic lateral sclerosis (ALS). The worm's simplicity and genetic tractability allow researchers to create models that express human disease-related proteins, such as amyloid-beta in AD or alpha-synuclein in PD, and study the resulting neurodegeneration [67].

## 8. *C. elegans* as animal model for neurodegenerative diseases

For nearly 50 years, the nematode *Caenorhabditis elegans* has been employed as an animal model. It is a perfect model for researching the neural pathways involved in neurodegenerative diseases since it has simple and primitive tissues and organs. Because *C. elegans* may imitate human diseases and maintain preserved neural circuits, it can provide important insights into the phenotypic of human diseases [68]. The relevance and application of *C. elegans* in Alzheimer's disease (AD), Parkinson's disease (PD), and Amyotrophic Lateral Sclerosis (ALS) are discussed below.

- Alzheimer's disease (AD)

  Research using *C. elegans* has provided significant insights into the molecular mechanisms underlying Alzheimer's disease. Studies have identified key genes and pathways involved in amyloid-beta aggregation and neurodegeneration, highlighting potential therapeutic targets [9]. The simplicity and genetic tractability of *C. elegans* make it an ideal model for high-throughput screening of anti-amyloid compounds [67]. Alzheimer's disease is characterized by the accumulation of amyloid-beta plaques and tau tangles in the brain, leading to neurodegeneration and cognitive decline. *C. elegans* models of AD have been created by overexpressing human amyloid-beta in the worm's neurons or muscle cells, leading to protein aggregation and neurotoxicity. These models have provided insights into the cellular processes that drive amyloid-beta aggregation, such as oxidative stress and mitochondrial dysfunction. Furthermore, *C. elegans* has been used to screen for compounds that can prevent or reverse amyloid-beta aggregation, identifying potential therapeutic candidates for AD.

- Parkinson's disease (PD)

  Parkinson's disease is caused by the loss of dopaminergic neurons in the brain, leading to motor deficits and tremors. C. elegans models of PD have been generated by overexpressing human alpha-synuclein, a protein that aggregates in the brains of PD patients [66]. These models exhibit neuronal degeneration and behavioral defects similar to those seen in PD patients, providing a platform for studying the molecular mechanisms of neurodegeneration. Studies in C. elegans have revealed that defects in mitochondrial function, autophagy, and lysosomal degradation contribute to alpha-synuclein toxicity, identifying potential targets for therapeutic intervention. C. elegans models of Parkinson's disease have been instrumental in elucidating the role of alpha-synuclein and mitochondrial dysfunction in neurodegeneration. These studies have identified genetic and environmental factors that contribute to disease progression, providing a foundation for the development of novel therapeutic strategies [69]. The conservation of key molecular pathways between C. elegans and humans underscores the translational potential of these findings [70].

- Amyotrophic lateral sclerosis (ALS)

  ALS is a neurodegenerative disease characterized by the progressive loss of motor neurons, leading to muscle weakness and paralysis. Mutations in the superoxide dismutase 1 (SOD1) gene are a common cause of familial ALS, and C. elegans models expressing mutant SOD1 have been used to study the disease's pathogenesis. These models exhibit motor neuron degeneration, mitochondrial dysfunction, and protein aggregation, providing insights into the cellular processes that drive ALS. Furthermore, C. elegans has been used to screen for compounds that can mitigate SOD1 toxicity and protect motor neurons, identifying potential therapeutic strategies for ALS [71].

## 9. Challenges and limitations

*C. elegans* has valuable features for aging studies, but it lacks certain anatomical characteristics found in mammals, such as a blood transport system and a blood-brain barrier, which may limit its applicability to human research [72]. Additionally, the absence of DNA methylation and long-range transcriptional regulation restricts its ability to model complex mechanisms relevant to other animal species, though it remains a useful simplified model for signal mapping studies [73].

Moreover, transgenes in this worm are found as extrachromosomal arrays and not in the genome; thus, the level of overexpression may be higher than what is found *in vivo*. Most neurodegenerative diseases are a product of multiple factors. However, the nervous system of *C. elegans* is simple compared to that of humans in terms of involvement and response to multifactorial risk contributing factors to the pathogenesis of the disease. The neural connectome of the human is more complex than that of the nematode and this may be a contributing factor to the expression of the many nervous diseases. However, several pathways and signaling molecules in this novel model are still conserved between the worms and mammals.

Overexpression of α-syn in *Caenorhabditis elegans* leads to age-independent motor defects and neuron loss, with aggregates resembling those found in human PD neurons. It has been observed that the absence of tdo-2 in *C. elegans*, which encodes the TDO-2 (tryptophan 2,3-dioxygenase), results in extended lifespan, reduced α-synuclein toxicity and increased tryptophan levels [74]. This suggests that TDO-2 regulates tryptophan-dependent toxicity. Furthermore, an RNAi screen in worms expressing α-synuclein in dopaminergic neurons identified five genes with potential neuroprotective functions, most of which are associated with vesicle trafficking [75]. These findings, along with the neuroprotective effects of TOR-2 and mammalian Rab1A, a GTPase involved in ER-to-golgi transport, point to a link between impaired ER-to-golgi vesicular transport and α-synuclein toxicity [76].

## 10. Future prospects

*C. elegans* has significantly advanced our understanding of neurodevelopmental disorders by revealing the genetic and cellular mechanisms that govern neural development. Research has uncovered critical genes and signaling pathways involved in axon guidance, synapse formation, and neuronal differentiation [77]. These findings have important implications for understanding and treating conditions such as autism and intellectual disability [78].

*C. elegans* has become an indispensable tool in translational neurobiology research, offering unique advantages for studying the molecular and cellular mechanisms of neurological diseases. The worm's simple nervous system, genetic tractability, and conservation of key molecular pathways make it an ideal model for investigating the genetic and environmental factors that contribute to neurodegeneration. Furthermore, *C. elegans* has been instrumental in identifying potential therapeutic targets and screening for drugs that can ameliorate neurodegenerative diseases.

As new technologies, such as CRISPR-Cas9, optogenetics, and advanced imaging techniques, continue to evolve, the potential of *C. elegans* in neurobiology research will only grow. Future studies will likely focus on integrating these technologies with high-throughput screening methods to accelerate the discovery of novel therapeutic compounds for human neurological disorders. Additionally, the use of *C. elegans* in aging research will provide valuable insights into how the nervous system declines with age and how this process can be slowed or reversed.

## 11. Conclusion

In conclusion, *C. elegans* remains a powerful and versatile model organism for translational neurobiology research, offering valuable insights into the mechanisms of neurodegeneration and the development of therapeutic strategies for neurological diseases.

Neurodegenerative diseases pose a significant and increasing global burden, with the need for extensive research to identify factors that mitigate their impact. While rodent models have advanced our understanding of these neurodegenerative and neurodevelopmental disorders, the phenotypic variations in humans suggest that utilizing a diverse range of model organisms is essential for accelerating research into their causes and potential cures [51]. Despite its evolutionary simplicity, *C. elegans* possesses numerous conserved cellular mechanisms and molecular pathways that are also found in mammals.

*C. elegans* shares many conserved molecular pathways and cellular mechanisms with mammals, making it a valuable model for comparative studies; tools like OrthoList 2 highlight the presence of 7943 genes, or approximately 41% of the *C. elegans* protein-coding genome, that have human orthologs [79]. The adult *C. elegans* (L4) hermaphrodite possesses about 300 neurons compared to the billion found in the human brain. This reduces complexity and the conservation of key components of synaptic transmission, including the neurotransmitters and receptors, making it a great tool in neurobiology [80, 81].

The worm has a great application in a neurobiological study involving Parkinson's disease, frontotemporal dementia, Huntington's disease, Huntington's disease, autism, and ALS. It is also applicable to understanding the interaction of the genetic-environmental factors in their pathogenesis and progression.

## Author details

Gideon Olamilekan Oluwatunase[1*], Patrick Godwin Okwute[2], Olayinka Olawale Asafa[3], David Anuoluwapo Oyeniran[1] and Onome Bright Ogenetega[2]

1 Department of Anatomy, Faculty of Basic Medical Sciences, University of Medical Sciences, Ondo City, Ondo State, Nigeria

2 Department of Physiology, Faculty of Basic Medical Sciences, Babcock University, Ilisan-Remo, Ogun State, Nigeria

3 Department of Physiology, Faculty of Basic Medical Sciences, Anchor University, Lagos State, Nigeria

*Address all correspondence to: goluwatunase@unimed.edu.ng

## References

[1] Davies C, Hamilton OKL, Hooley M, Ritakari TE, Stevenson AJ, Wheater ENW. Translational neuroscience: The state of the nation (a PhD student perspective). Brain Communications. 2020;**2**(1):fcaa038

[2] Riddle DL, Blumenthal T, Meyer BJ, Priess JR. *C. elegans* II. 2nd ed. Cold Spring Harbor (NY): Cold Spring Harbor Laboratory Press; 1997

[3] Shen P, Yue Y, Park Y. A living model for obesity and aging research: *Caenorhabditis elegans*. Critical Reviews in Food Science and Nutrition. 2018;**58**(5):741-754

[4] Hall DH, Altun ZF. *C. elegans* Atlas. Cold Spring Harbor Laboratory Press. UK: Cambridge University Press; 2008. ISBN 978-087969715-0

[5] Harrington AJ, Hamamichi S, Caldwell GA, Caldwell KA. *C. elegans* as a model organism to investigate molecular pathways involved with Parkinson's disease. Developmental Dynamics. 2010;**239**(5):1282-1295

[6] Brenner S. The genetics of *Caenorhabditis elegans*. Genetics. 1974;**77**(1):71-94

[7] Giunti S, Anderson N, Rayes D, De Rosa MJ. Drug discovery: Insights from the invertebrate *Caenorhabditis elegans*. Pharmacology Research & Perspectives. 2021;**9**(2):e00721. DOI: 10.1002/prp2.721

[8] Leung MC, Williams PL, Benedetto A, Au C, Helmcke KJ, Aschner M, et al. *Caenorhabditis elegans*: An emerging model in biomedical and environmental toxicology. Toxicological Sciences. 2008;**106**(1):5-28

[9] Kaletta T, Hengartner MO. Finding function in novel targets: *C. elegans* as a model organism. Nature Reviews. Drug Discovery. 2006;**5**(5):387-399

[10] Apfeld J, Alper S. What can we learn about human disease from the Nematode C. *elegans*? Methods in Molecular Biology. 2018;**1706**:53-75

[11] Ijomone OM, Weishaupt A-K, Michaelis V, Ijomone OK, Bornhorst J. p38- and ERK-MAPK signalling modulate developmental neurotoxicity of nickel and vanadium in the *Caenorhabditis elegans* model. Kinases Phosphatases. 2024;**2**(1):28-42

[12] Cassada RC, Russell RL. The dauerlarva, a post-embryonic developmental variant of the nematode *Caenorhabditis elegans*. Developmental Biology. 1975;**46**(2):326-342

[13] Chen Y, Scarcelli V, Legouis R. Approaches for studying autophagy in *Caenorhabditis elegans*. Cells. 2017;**6**(3):27

[14] Schaffitzel E, Hertweck M. Recent aging research in *Caenorhabditis elegans*. Experimental Gerontology. 2006;**41**(6):557-563

[15] Matsunami K. Frailty and *Caenorhabditis elegans* as a benchtop animal model for screening drugs including natural herbs. Frontiers in Nutrition. 2018;**5**:111

[16] David DC, Ollikainen N, Trinidad JC, Cary MP, Burlingame AL, Kenyon C. Widespread protein aggregation as an inherent part of aging in *C. elegans*. PLoS Biology. 2010;**8**(8):e1000450

[17] Zhang S, Li F, Zhou T, Wang G, Li Z. *Caenorhabditis elegans* as a useful

model for studying aging mutations. Frontiers in Endocrinology (Lausanne). 2020;**11**:554994. DOI: 10.3389/fendo.2020.554994

[18] Hertweck M, Hoppe T, Baumeister R. *C. elegans*, a model for aging with high-throughput capacity. Experimental Gerontology. 2003;**38**(3):345-346

[19] Kumar J, Park K-C, Awasthi A, Prasad B. Silymarin extends lifespan and reduces proteotoxicity in *C. elegans* Alzheimer's model. CNS & Neurological Disorders Drug Targets. 2015;**14**(2):295-302

[20] Cong W, Wang P, Qu Y, Tang J, Bai R, Zhao Y, et al. Evaluation of the influence of fullerenol on aging and stress resistance using *Caenorhabditis elegans*. Biomaterials. 2015;**42**:78-86

[21] White JG, Southgate E, Thomson JN, Brenner S. The structure of the nervous system of the nematode *Caenorhabditis elegans*. Philosophical Transactions of the Royal Society of London. Series B, Biological Sciences. 1986;**314**(1165):1-340

[22] Chalfie M, Tu Y, Euskirchen G, Ward WW, Prasher DC. Green fluorescent protein as a marker for gene expression. Science. 1994;**263**(5148):802-805

[23] Queirós L, Marques C, Pereira JL, Gonçalves FJM, Aschner M, Pereira P. Overview of chemotaxis behavior assays in *Caenorhabditis elegans*. Current Protocols. 2021;**1**(5):e120. DOI: 10.1002/cpz1.120

[24] Alexander AG, Marfil V, Li C. Use of *C. elegans* as a model to study Alzheimer's disease and other neurodegenerative diseases. Frontiers in Genetics. 2014;**5**:279

[25] Harrington AJ, Yacoubian TA, Slone SR, Caldwell KA, Caldwell GA. Functional analysis of VPS41-mediated neuroprotection in *Caenorhabditis elegans* and mammalian models of Parkinson's disease. The Journal of Neuroscience. 2018;**32**(6):2142-2153

[26] Zhang Y, Mair WB. *C. elegans* as a model for studying the molecular mechanisms of aging and neurodegeneration. Molecular Neurobiology. 2019;**56**(7):4656-4673

[27] MacDonald ME, Ambrose CM, Duyao MP, Myers RH, Lin C, Srinidhi L, et al. A novel gene containing a trinucleotide repeat that is expanded and unstable on Huntington's disease chromosomes. Cell. 1993;**72**(6):971-983. DOI: 10.1016/0092-8674(93)90585-E

[28] Roussos A, Kitopoulou K, Borbolis F, Palikaras K. *Caenorhabditis elegans* as a model system to study human neurodegenerative disorders. Biomolecules. 2023;**13**(3):478. DOI: 10.3390/biom13030478

[29] Faber PW, Alter JR, MacDonald ME, Hart AC. Polyglutamine-mediated dysfunction and apoptotic death of a *Caenorhabditis elegans* sensory neuron. Proceedings of the National Academy of Sciences of the United States of America. 1999;**96**(1):179-184. DOI: 10.1073/pnas.96.1.179

[30] Parker JA, Holbert S, Lambert E, Abderrahmane S, Néri C. Genetic and pharmacological suppression of polyglutamine-dependent neuronal dysfunction in *Caenorhabditis elegans*. Journal of Molecular Neuroscience. 2004;**23**:61-67

[31] Kourtis N, Nikoletopoulou V, Tavernarakis N. Small heat-shock proteins protect from heat-stroke-associated neurodegeneration. Nature.

2012;**490**(7419):213-218. DOI: 10.1038/nature11417

[32] Cordeiro LM, Machado ML, da Silva AF, Baptista FBO, da Silveira TL, Soares FAA, et al. Rutin protects Huntington's disease through the insulin/IGF1 (IIS) signaling pathway and autophagy activity: Study in *Caenorhabditis elegans* model. Food and Chemical Toxicology. 2020;**141**:111323

[33] Cordeiro LM, Soares MV, da Silva AF, Machado ML, Baptista FBO, da Silveira TL, et al. Neuroprotective effects of rutin on ASH neurons in *Caenorhabditis elegans* model of Huntington's disease. Nutritional Neuroscience. 2022;**25**(11):2288-2301

[34] Rapti G, Richmond JE, Bessereau JL. A single immunoglobulin-domain protein required for clustering acetylcholine receptors in *C. elegans*. The EMBO Journal. 2017;**36**(3):225-239

[35] Sala AJ, Pielage J. Protein trafficking in synapse development and plasticity in *C. elegans* and drosophila. Cellular and Molecular Life Sciences. 2020;**77**(18):3681-3700

[36] Piggott BJ, Liu J, Feng Z, Wescott SA, Xu XZS. The neural circuits and synaptic mechanisms underlying motor initiation in *C. elegans*. Cell. 2011;**147**(4):922-933

[37] Ardiel EL, Rankin CH. An elegant mind: Learning and memory in *C. elegans*. Learning & Memory. 2010;**17**(4):191-201

[38] Goodman MB, Sengupta P. How C. *elegans* senses mechanical stress, temperature, and other physical stimuli. Genetics. 2019;**212**(1):25-51

[39] Leung MC, Goldstone JV, Boyd WA, Freedman JH, Meyer JN. *Caenorhabditis elegans* research community. *C. elegans* generates biologically relevant levels of reactive oxygen species that are detectable using a hydrogen peroxide-specific sensor. Nature Methods. 2008;**5**(4):281-282

[40] Davis CA et al. CRISPR-Cas9-mediated gene editing in *C. elegans*: A versatile tool for neurobiological research. Journal of Neuroscience Methods. 2022;**300**:101-110

[41] Crombie TA et al. CaeNDR, the Caenorhabditis natural diversity resource. Nucleic Acids Research. 2023;**52**(D1):D850-D858

[42] Crombie TA et al. RNA interference in *C. elegans*: Insights into gene function and disease models. Molecular Genetics and Genomics. 2023;**598**:204-220

[43] Frézal L, Félix MA. The natural history of model organisms: *C. elegans*. Nature Reviews. Genetics. 2015;**16**(9):605-615

[44] Sengupta P, Samuel ADT. *Caenorhabditis elegans*: A model system for systems neuroscience. Current Opinion in Neurobiology. 2009;**19**(1):1-7

[45] Kerr R et al. Calcium imaging in *C. elegans*: A window into neural activity. Neuron. 2000;**26**(2):303-314

[46] Liang YL et al. Neuroscience: Electrophysiology in intact *Caenorhabditis elegans*. Nature. 2017:118-123

[47] Nectow AR et al. Building a better TRAP for translation. Nature Methods. 2017;**14**(10):1021-1028

[48] Gottschalk A, Almedom R. Electrophysiological techniques in *C. elegans* neurobiology research: Challenges and breakthroughs. Trends in Neurosciences. 2004;**24**(12):689-695

[49] Fryer E et al. A high-throughput behavioral screening platform for measuring chemotaxis by *C. elegans*. PLoS Biology. 2024;**22**(6):e3002672

[50] Thompson M et al. A *Caenorhabditis elegans* behavioral assay distinguishes early-stage prostate cancer patient urine from controls. Biology Open. 2021;**10**(3):bio057398

[51] Caldwell KA, Willicott CW, Caldwell GA. Modeling neurodegeneration in *Caenorhabditis elegans*. Disease Models & Mechanisms. 2020;**13**(10):dmm046110. DOI: 10.1242/dmm.046110

[52] Hardy J, Selkoe DJ. The amyloid hypothesis of Alzheimer's disease: Progress and problems on the road to therapeutics. Science. 2002;**297**(5580):353-356

[53] Palop JJ, Mucke L. Epilepsy and cognitive impairments in Alzheimer disease. Archives of Neurology. 2009;**66**(4):435-440

[54] Treusch S, Hamamichi S, Goodman JL, Matlack KE, Chung CY, Baru V, et al. Functional links between Aβ toxicity, endocytic trafficking, and Alzheimer's disease risk factors in yeast. Science. 2011;**334**(6060):1241-1245

[55] Cherny RA, Atwood CS, Xilinas ME, Gray DN, Jones WD, McLean CA, et al. Treatment with a copper-zinc chelator markedly and rapidly inhibits β-amyloid accumulation in Alzheimer's disease transgenic mice. Neuron. 2001;**30**(3):665-676

[56] Matlack KE, Tardiff DF, Narayan P, Hamamichi S, Caldwell KA, Caldwell GA, et al. Clioquinol promotes the degradation of metal-dependent amyloid-β (Aβ) oligomers to restore endocytosis and ameliorate Aβ toxicity. National Academy of Sciences of the United States of America. 2014;**111**(11):4013-4018

[57] Griffin EF, Scopel SE, Stephen CA, Holzhauer AC, Vaji MA, Tuckey RA, et al. ApoE-associated modulation of neuroprotection from Aβ-mediated neurodegeneration in transgenic *Caenorhabditis elegans*. Disease Models & Mechanisms. 2019;**12**(2):dmm037218

[58] Spinney L. Alzheimer's disease: The forgetting gene. Nature. 2014;**510**:7503

[59] Jankovic J, Tan EK. Parkinson's disease: Etiopathogenesis and treatment. Journal of Neurology, Neurosurgery, and Psychiatry. 2020;**91**(8):795-808

[60] Poewe W, Seppi K, Tanner CM, Halliday GM, Brundin P, Volkmann J, et al. Parkinson disease. Nature Reviews. Disease Primers. 2017;**3**(1):1-21

[61] Ijomone OM, Ijomone OK, Iroegbu JD, Ifenatuoha CW, Olung NF, Aschner M. Epigenetic influence of environmentally neurotoxic metals. Neurotoxicology. 2020;**81**:51-65. DOI: 10.1016/j.neuro.2020.08.005

[62] Mocko JB, Kern A, Moosmann B, Behl C, Hajieva P. Phenothiazines interfere with dopaminergic neurodegeneration in *Caenorhabditis elegans* models of Parkinson's disease. Neurobiology of Disease. 2010;**40**(1):120-129

[63] Spillantini MG, Schmidt ML, Lee VM-Y, Trojanowski JQ, Jakes R, Goedert M. α-Synuclein in Lewy bodies. Nature. 1997;**388**(6645):839-840

[64] Goedert M. Alpha-synuclein and neurodegenerative diseases. Nature Reviews. Neuroscience. 2001;**2**(7):492-501

[65] Stefanis L. α-Synuclein in Parkinson's disease. Cold Spring Harbor Perspectives

in Medicine. 2012;**2**(2):a009399. DOI: 10.1101/cshperspect.a009399

[66] Lakso M, Vartiainen S, Moilanen AM, Sirviö J, Thomas JH, Nass R, et al. Dopaminergic neuronal loss and motor deficits in *Caenorhabditis elegans* overexpressing human α-synuclein. Journal of Neurochemistry. 2003;**86**(1):165-172

[67] Benedetto A et al. *Caenorhabditis elegans*: An emerging model in biomedical and environmental toxicology. Toxicological Sciences. 2008;**106**(1):5-28

[68] Caldero-Escudero E, Romero-Sanz S, De la Fuente S. Using C. *elegans* as a model for neurodegenerative diseases: Methodology and evaluation. Methods in Cell Biology. 2024;**188**:1-34

[69] National Research Council (NRC). Toxicity Testing in the 21st Century: A Vision and a Strategy. Washington DC: National Academies Press; 2000

[70] Hope IA. *C. elegans*: A Practical Approach. UK: Oxford University Press; 1999

[71] Masrori P, Van Damme P. Amyotrophic lateral sclerosis: A clinical review. European Journal of Neurology. 2020;**27**(10):1918-1929. DOI: 10.1111/ene.14393. Epub 2020 Jul 7

[72] Weinhouse C, Truong L, Meyer JN, Allard P. *Caenorhabditis elegans* as an emerging model system in environmental epigenetics. Environmental and Molecular Mutagenesis. 2018;**59**(7):560-575

[73] Corsi AK, Wightman B, Chalfie M. A transparent window into biology: A primer on *Caenorhabditis elegans*. Genetics. 2015;**200**(2):387-407

[74] Van der Goot AT, Zhu W, Vázquez-Manrique RP, Seinstra RI, Dettmer K, Michels H, et al. Delaying aging and the aging-associated decline in protein homeostasis by inhibition of tryptophan degradation. Proceedings of the National Academy of Sciences of the United States of America. 2012;**109**(37):14912-14917. DOI: 10.1073/pnas.1203083109

[75] Hamamichi S, Rivas RN, Knight AL, Cao S, Caldwell KA, Caldwell GA. Hypothesis-based RNAi screening identifies neuroprotective genes in a Parkinson's disease model. Proceedings of the National Academy of Sciences of the United States of America. 2008;**105**(2):728-733. DOI: 10.1073/pnas.0711018105

[76] Cooper AA, Gitler AD, Cashikar A, Haynes CM, Hill KJ, Bhullar B, et al. Alpha-synuclein blocks ER-Golgi traffic and Rab1 rescues neuron loss in Parkinson's models. Science. 2006;**313**(5785):324-328. DOI: 10.1126/science.1129462

[77] Antoshechkin I, Sternberg PW. The C. *elegans* Research Community. WormBook; 2007 (Online database issue)

[78] Frézal L, Félix M. The natural history of model organisms: *C. elegans* outside the petri dish. eLife. 2005;**4**:e05849

[79] Shaye DD, Greenwald I. OrthoList: A compendium of *C. elegans* genes with human orthologs. PLoS One. 2011;**6**(5):e20085

[80] Cook SJ, Jarrell TA, Brittin CA, Wang Y, Bloniarz AE, Yakovlev MA, et al. Whole-animal connectomes of both *Caenorhabditis elegans* sexes. Nature. 2019;**571**(7763):63-71

[81] Li C, Kim K. Neuropeptides. In: WormBook: The Online Review of *C. elegans* Biology. California: eScholarship; 2008. pp. 1-36. DOI: 10.1895/wormbook.1.142.1

Chapter 4

# Animal Models in Translational Pain Research

*Zhichao Wang, Lumei Xiang, Xiaoxia Wang, Xiaoqing Tan and Anfeng Xiang*

## Abstract

Animal models play a crucial role in translational pain research. They provide a means to understand the underlying mechanisms of pain, test potential therapeutic approaches, and simulate various pain conditions. This chapter explores the significance and application of different animal models in this field. Rodent models, such as mice and rats, are commonly used due to their genetic manipulability and ease of handling. For instance, neuropathic pain models can be induced by nerve injury to study the changes in neuronal circuitry and molecular pathways. Moreover, primate models offer closer similarities to human physiology and behavior, allowing for more accurate assessment of pain perception and responses. However, each model has its limitations, and careful consideration must be given to the selection and interpretation of results. Future research should focus on developing more refined and clinically relevant animal models to accelerate the translation of findings from bench to bedside and improve pain management strategies for patients.

**Keywords:** pain management, animal models, pain model, nervous system, life science

## 1. Introduction

Animal models are essential in life sciences for research and drug development. Advances in gene-editing technologies, such as CRISPR-Cas9, have enabled more precise genetic modifications. This precision allows for the creation of animal models that accurately mimic human diseases, thereby speeding up drug discovery and the study of disease mechanisms. However, these advancements have also brought ethical and animal welfare concerns to the forefront. As a result, there is a growing movement toward developing ethical alternatives and reducing the use of experimental animals. The 3R principle—replacement, reduction, and refinement—is expected to become increasingly important to ensure that animal experiments are necessary and that animal suffering is minimized. Furthermore, standardizing and sharing animal model resources will likely improve scientific collaboration and efficiency.

## 2. Implications and challenges in pain research

### 2.1 Definition of pain

Pain is an unpleasant sensory and emotional experience linked to actual or potential tissue damage [1]. The perception of pain varies greatly among individuals. Even when experiencing the same injury, people may feel pain differently. Because pain is a highly personal experience, self-reports are the most reliable way to measure it. Even so, individual differences can cause the same stimulus to elicit different responses. Pain sensations can range from mild to severe and may include tingling, burning, shooting, or electrical shock-like feelings.

### 2.2 The significance of pain

Pain functions as a vital survival mechanism by alerting us to harmful changes in the body, such as the presence of cancer. It also serves as a warning system, teaching us to avoid harmful stimuli, like touching hot surfaces. In this way, specific types of pain can protect us from injury. However, sometimes pain continues without any clear benefit, indicating a need for further understanding and management.

### 2.3 Physiological basis of pain

Pain is a complex neural process that involves both the peripheral and central nervous systems. Various regions of the central nervous system, including the insular cortex, somatosensory cortex, hippocampus, prefrontal cortex, motor cortex, basal ganglia, anterior cingulate cortex (ACC), spinal cord, amygdala, brainstem, midbrain, and thalamus, are involved in processing pain. This indicates that pain is not solely a physical sensation but also encompasses emotional aspects. Biological, psychological, and social factors all interact to influence the experience and management of both acute and chronic pain [2, 3]. Psychological factors, in particular, have received increasing attention for their role in influencing pain perception. In addition, a significant link between pain and sleep has been observed. In clinical practice, it is common to find that patients suffering from long-term chronic pain often experience disruptions in sleep. A study published in *Neuron*, titled "Anterior cingulate cortex projections to the dorsal medial striatum underlie insomnia associated with chronic pain" revealed that the anterior cingulate cortex–dorsal medial striatum (ACC–DMS) pathway plays a key role in pain-related insomnia, thus providing a solid foundation for us to explore the relationship between pain and sleep.

### 2.4 Research progress on pain

In the frontier research of pain science, the emphasis has been widely placed on understanding and simulating acute, chronic, or persistent pain in clinical patients. Pain model development methods vary and include chemical irritation models, physical irritation models, neurogenic injury models, and visceral pain models, among others. An ideal pain model should meet the following basic requirements: (1) The cause of pain should resemble the clinical situation; (2) The animal's pain-related behavioral response and reflexive escape responses should be objectively detectable and similar to chronic pain symptoms; (3) The model should facilitate the study of pain mechanisms; and (4) The model should be simple to prepare with a high success rate [4].

However, creating a model that reflects pain caused by multiple etiologies is challenging. The development and use of animal models aim to translate subjective assessments into objective and reproducible behavioral measures with predictive accuracy and reliability. This has led to a focus on evaluating preclinical efficacy in animal models, which can sometimes be limited by the methodologies used. Concerns have been raised regarding the reproducibility of results, publication bias, and inadequate preparation before clinical trials. These issues complicate the ability of animal models to accurately predict clinical efficacy [5].

Another concern is that outcome measures in animal studies often do not match with those in clinical trials. In humans, pain can be reported by individuals, but in animals, researchers rely on observing responses to putative sources of pain. For example, in human trials, the main efficacy indicators are usually spontaneous and sustained activity, including assessments of physical and emotional functioning and sleep disturbances, which are elements rarely evaluated in animal efficacy studies, making direct comparison difficult [6].

Despite these methodological and procedural concerns, animal models remain crucial for pain research. They help elucidate pain mechanisms and pathophysiology and provide guidance for developing new therapeutic goals and compounds for clinical assessment [7]. Continuous evaluation and improvement of existing models, as well as the development of new ones, are essential for advancing pain research.

### 2.5 Cutting-edge research methods on pain

The application of emerging technologies is transforming pain research. Neuroimaging techniques, such as functional magnetic resonance imaging (fMRI), electroencephalography (EEG), and magnetoencephalography (MEG), are instrumental in observing and analyzing transmission pathways and neural activity patterns of pain signals in the brain. These methods offer valuable insights into the neural mechanisms underlying pain [8].

In addition, advanced techniques like optogenetic technology, genomics, proteomics, and single-cell sequencing [9] are providing new avenues for understanding pain.

## 3. Significance and status of animal models in pain research

### 3.1 Animal models of human diseases

Animal models, which replicate characteristics of human diseases, are crucial in modern biomedical research. They facilitate a deeper understanding of disease onset and progression and aid in the research of prevention and control measures.

### 3.2 Superiority of animal models

Human-based research is constrained by ethical, moral, and methodological limitations. Animal models offer an effective solution to these challenges and have become an indispensable tool in biomedical research due to the following advantages.

### *3.2.1 Avoiding risks in human experiments*

Clinical research, especially on external injuries, poisoning, and inflammation, is difficult or even impossible to conduct on human subjects due to ethical and safety concerns. For example, in research on acute and chronic respiratory diseases, it is difficult to ignore the role of environmental pollution.

3.2.2 Replicating rare or complex clinical conditions. Some diseases, such as radiation sickness, gas poisoning, or severe infectious diseases, are rare or difficult to study in humans due to their low incidence. However, animal models allow researchers to induce these conditions at will, facilitating timely investigation.

### *3.2.2 Overcoming long latency and low incidence of diseases*

Certain diseases in humans, such as acute leukemia or autoimmune disorders, have low incidences or long latency periods. Animal models help accelerate research by enabling higher frequencies of disease induction, allowing scientists to study the disease mechanisms and progression in a controlled environment.

### *3.2.3 Control of experimental conditions*

Many human diseases are very complex and often exacerbated by overlapping conditions. For example, patients with heart disease may also suffer from lung or kidney diseases. Animal models allow researchers to control variables such as genetics, age, sex, and environmental factors, leading to more consistent and comparable experimental results.

### *3.2.4 Simplified sampling and experimental procedures*

Animal models act as "microcosms" of human diseases, allowing researchers to collect samples at any stage of the experiment. This level of control is often impossible in human studies. The trend toward miniaturized animal models enhances ease of management and experimental manipulation.

### *3.2.5 Comprehensive understanding of disease mechanisms*

Through the comparative study of zoonoses across species, researchers can gain a deeper understanding of how the same pathogen or condition affects different organisms. This comparative approach enriches the study of human diseases by revealing diverse pathological changes that may not be evident in clinical research alone.

Therefore, the use of animal disease models to study human diseases can overcome the challenges posed by the complexity, long incubation periods, and low incidence of human diseases. They allow for the replication of disease conditions with a single etiology in a controlled environment, making them invaluable tools in studying disease mechanisms and treatment interventions [10].

While animal research has contributed significantly to scientific and medical advancements, it is essential to prioritize animal welfare. The 3R principles should be systematically considered in all studies involving animals [11]. Although caring for research animals presents emotional challenges, it ultimately drives productivity in biomedical research [12].

## 4. Patterns and types of pain

According to the classification criteria proposed by the National Institute of Neurological Diseases and Stroke (NINDS), there are three major patterns of pain based on duration and frequency. (Pain | National Institute of Neurological Disorders and Stroke (nih.gov)).

### 4.1 Pain classification by frequency

#### *4.1.1 Acute pain*

Acute pain begins suddenly and ends when its etiology is treated or cured. Acute pain often serves as a warning sign that injury, illness, overuse, or other environmental stress pose a threat to the body. Common causes of acute pain are muscle strain, bone fractures, dental work, surgery, childbirth, infections, burns, and so on.

#### *4.1.2 Paroxysmal pain*

Paroxysmal pain occurs intermittently and may be associated with chronic conditions like sickle cell disease. Chronic migraine is another example of paroxysmal pain, which can be widespread or triggered by known factors.

#### *4.1.3 Chronic pain*

Chronic pain persists for extended periods, often defined as lasting longer than 12 weeks. It may evolve from acute pain conditions or occur without a clear cause. Individuals may experience one or more chronic pain conditions, or a combination of chronic and acute pain.

### 4.2 Pain classification by etiology

#### *4.2.1 Nociceptive pain*

Nociceptive pain results from tissue damage and/or inflammation, with sensations ranging from sharp to dull depending on the injury or inflammation. Examples include pain from paper cuts, infections, bone fractures, or osteoarthritis.

#### *4.2.2 Neuropathic pain*

Neuropathic pain arises from damage or dysfunction in the nervous system, resulting from injury or disease. Neuropathic pain sensations are often described as burning, tingling, shooting, or similar to electrical shocks. Diseases that cause neuropathic pain include diabetic neuropathy, herpes zoster, and sciatica.

#### *4.2.3 Nociplastic pain*

Nociplastic pain occurs due to altered nociception, where there is no clear evidence of tissue damage, inflammation, or nerve injury. This type of pain is closely related to sensory hypersensitivity. Examples of nociplastic pain conditions include fibromyalgia, irritable bowel syndrome, and chronic low back pain.

# 5. Classification and selection of animal models

## 5.1 Classification of animal models

### *5.1.1 Traditional laboratory animal classification*

Traditional experimental animals are primarily mammals of the vertebrate family. Commonly used species include mice, rats, guinea pigs, ground mice, rabbits, dogs, pigs, and monkeys. In recent years, animals such as the Oriental vole, tree shrew, and ferret have gained experimental significance due to their unique characteristics. In addition, nonmammalian animals like nematodes, fruit flies, silkworms, zebrafish, and *Xenopus* (African clawed frog) have become popular due to their affordability, ease of use, and clearly defined biology.

### *5.1.2 Genetic classification of experimental animals*

Experimental animals can be classified based on their genetic characteristics into three main groups: inbred, closed, and hybrid.

### *5.1.3 Microbiological and parasitological classification of experimental animals*

Based on the level of control over microorganisms and parasites in or on the animal, laboratory animals are classified into four categories: general grade animals, clean animals, specific pathogen-free (SPF) animals, and germ-free animals [13].

Conventional animal (CV) refers to pathogens that do not carry major zoonotic pathogens and severe animal infectious diseases. The public believes that ordinary animals are more suitable for teaching and are generally not used for scientific research experiments. Pathogens that should be excluded include: ectoparasites, dermatophytes, salmonella, lymphocytic choriomeningitis virus, epidemic hemorrhagic fever virus, mousepox virus, etc. In addition to the pathogens that should be eliminated for ordinary animals, clean animal (CL) is required not to carry pathogens that are harmful to animals and interfere with scientific research. It can be applied to short-term experiments and some scientific experiments. In addition to the requirements for ordinary animals, clean animal (CL) should also be excluded. Pathogens include: most internal parasites, Taizeella, *Corynebacterium muris*, *Bordetella bronchiseptica*, murine hepatitis virus, Sendai virus, etc. Specific pathogen free animal (SPF) In addition to the pathogens that should be excluded in first- and second-level animals, it cannot carry pathogens that are major potential infections or conditional pathogens and interfere with scientific experiments, such as flagellate parasites and *Staphylococcus aureus*, *Klebsiella* pneumoniae, *Streptococcus* pneumoniae, beta-hemolytic *Streptococcus*, *Pseudomonas aeruginosa*, pneumonia virus, parvovirus and 11 other viruses. SPF meets the standard level requirements of laboratory animal models and can be used for currently existing All science experiments. No microorganisms can be detected in germ free animals (GF), and they need to be cultured in isolation. They are often used in experimental research with higher cleanliness levels, such as the study of intestinal microflora [14].

## 5.2 Selection of animal models

The proper selection of experimental animals is a prerequisite for the success of biomedical research. Choosing the appropriate animal model is a key step, as different animals possess distinct biological, anatomical, and physiological characteristics that can influence the results of the experiment. Using the wrong animal model can lead to incorrect conclusions and potentially invalidate the entire experiment. Therefore, it is essential to choose experimental animals that maximize efficiency, using the least amount of human and material resources as well as time to achieve satisfactory results and experimental efficiency [15].

Selection of experimental animals for scientific research should be guided by the purpose and requirements of the experiment and whether the animals are easy to obtain, economical, and manageable in terms of housing, care, and experimental handling. Three factors must be considered when selecting experimental animals: species, breed, and strain. In addition, the following should also be given attention:

1. Similarity to humans: Select animals that closely mirror human functions, metabolism, structure, and disease characteristics.

2. Genetic background and stability: Animals with a clear genetic background and stable model traits should be used. Their microbiome and health status should also be known.

3. Anatomical and physiological fit: Selected animals should have anatomical and physiological traits that meet the requirements of the experiment.

4. Special responses in different strains: Select experimental animals that exhibit specific responses to the experimental intervention.

5. Zoonotic disease models and traditional laboratory animals: Animals that model human zoonotic diseases and those commonly used in past studies should be considered.

The selection of appropriate experimental animals directly affects the progression of biomedical research and the authenticity and accuracy of the results. Researchers must carefully balance selecting qualified, standardized animal models with practical considerations such as resource availability and national guidelines. The adoption of new technologies and research methods should also be encouraged to ensure the representativeness and relevance of animal models in biomedical research.

## 6. Overview of research progress in animal models of pain

### 6.1 Problems in pain research

Pain is recognized as the fifth vital sign in modern medicine, making pain management a priority in clinical practice. Pain can be classified into acute and chronic pain based on the duration of the injury [16]. Pain can lead to sleep disorders, appetite loss, mental breakdown, and even personality changes. In severe cases, prolonged pain can drive patients to consider suicide, placing a huge burden on families and society. Many scholars have developed animal models to better understand the mechanisms of pain and to guide the treatment and care of pain patients.

While many animal models have shown pathological and symptomatic manifestations similar to clinical situations, some limitations remain. Zhang and Luo [17] point out that there is still a gap between existing animal models and natural conditions. Current models primarily measure pain behavior by focusing on responses to temperature or mechanical stimulation, which do not fully align with the human experience of pain. In addition, researchers often focus only on pain itself, neglecting accompanying symptoms. Some studies have identified a connection between pain patterns and emotions such as anxiety and depression, supporting the notion that "pain is an unpleasant experience." This interaction between pain and emotional factors warrants further investigation.

## 6.2 Challenges in pain research

There is a significant mismatch between the epidemiological characteristics of chronic pain in the population and the characteristics selected in animal studies. Gender, age, and genetic factors strongly influence experimental results, yet there is no common consensus on how to standardize these variables [18, 19]. To better understand these impacts, we must examine the main methods of animal pain testing.

### *6.2.1 Overemphasis on reflexive responses*

Pain measurements often rely too heavily on reflexive withdrawal from mechanical or thermal (both hot and cold) stimuli as a dependent measure.

### *6.2.2 Neglecting associated conditions*

By concentrating exclusively on pain measurement, many researchers overlook associated conditions, including sequelae and comorbidities. Patients with chronic pain frequently suffer from disability, anxiety, depression, cognitive impairments, sleep disturbances, reduced libido, social withdrawal, and other issues. While these conditions can be replicated in animal models with chronic inflammatory or neuropathic states, the outcomes are often inconclusive.

### *6.2.3 Mismatch between human and animal study populations*

There is usually a clear mismatch between the epidemiological reality of the prevalence of chronic pain in the human population and the selection of animal models. Most animal models use young male or female mice due to convenience, but most chronic pain patients are female, and the prevalence of chronic pain is higher in the elderly.

### *6.2.4 Lack of robust experimental design*

The design and reporting standards of animal experiments often fall short compared to human clinical trials. Essential details such as blinding, randomization, and data monitoring are frequently omitted, leading to experimental bias. Additionally, many animal models are based on Western medical pathology, overlooking Traditional Chinese Medicine (TCM) and other theoretical frameworks. Developing TCM "syndrome" pain models in accordance with TCM principles can more effectively uncover the mechanisms of analgesia and pain pathogenesis [4].

Current animal models cannot fully replicate clinical pain conditions, such as phantom limb pain, headaches, or pain from spinal cord injuries, based solely on observing behavioral changes [20]. Focusing on cortical plasticity may address these limitations and provide a more comprehensive understanding.

Despite the obvious limitations of current animal pain models, significant progress has been made in understanding pain mechanisms and improving treatment strategies. With continued refinement and innovation, we believe that future animal pain models will be more likely to improve the treatment of human pain.

## 7. Advances in animal models of chronic pain

Chronic pain is defined as pain lasting more than 3 months. A quarter of the global population is plagued by chronic pain, leading to significant economic burden on both families and society [21, 22]. At present, the pathogenesis of chronic pain is unknown, the treatment measures for chronic pain are relatively limited, and the progress in efficacy has been slow [23]. In recent years, numerous animal studies have tried to clarify the pathogenesis of chronic pain through animal experiments and to further explore effective therapeutic measures [24]. This chapter explores the preparation of animal models of chronic pain and the relevant treatment options, providing a useful reference for the selection and establishment of animal models in chronic pain research.

### 7.1 Preparation of the animal models of chronic pain

Chronic pain models can be developed using various experimental animals. Among them, primates are considered ideal due to their highly similar anatomical structures and physiological functions to human. However, they are rarely studied in this context [23] because of high costs, stringent feeding requirements, and their limited availability. Large and medium-sized animals such as pigs, dogs, and cats have been used to study chronic pain but face similar limitations to primates, preventing widespread use. Rodents, on the other hand, are the most widely used in chronic pain studies [19] because pain significantly affects their behavior [25, 26], and they offer advantages of lower cost, ease of handling, and general anatomical and physiological similarity to humans. Commonly used rodents include rats, mice, and rabbits [19].

### 7.2 Animal models of chronic primary pain

Chronic pain can be classified in various ways. Based on the work of Professor Rolf-Detlef Treede [27], this section outlines the preparation of animal models for two main categories of chronic pain: chronic primary pain and chronic secondary pain. Some studies have defined chronic primary pain as the presence of persistent or recurrent pain for more than 3 months, with obvious emotional problems or dysfunction, without being attributed to other causes [28]. This includes conditions [27] like fibromyalgia, chronic primary cephalic-facial pain, chronic primary visceral pain, and chronic primary musculoskeletal pain.

#### *7.2.1 Chronic fibromyalgia*

Fibromyalgia is a chronic diffuse pain syndrome caused by dysfunctional processing of central nervous sensory afferents. Its main symptoms include multisite pain, severe fatigue, stiffness, sleep disturbances, cognitive impairment, and psychological problems [29, 30]. At present, the etiology and pathogenesis of fibromyalgia are still unclear [31]. In some studies, fibromyalgia models have been developed by anesthetizing mice and injecting 20 μL of acidic saline (pH 4.0) into the right gastrocnemius muscle twice daily, which establishes a fibromyalgia-like model. Mechanosensitivity is then tested by observing the mouse's response to stimulation with electronic filaments [32]. This preparation method is relatively simple and reproducible.

Another study [33] exposed mice to alternating cold and warm air, with cage temperatures ranging from 22 to 7°C at 30-minute intervals starting at 10:00 am to minimize the impact on their sleep cycles. Over seven consecutive days, repeated cold air stimulation was applied, and the skin tissue was examined to ensure no damage, inflammation, or edema occurred at the lower temperature of 7°C. Afterward, pain behavior was assessed by measuring the mechanical hind paw retraction threshold. Pressure was applied to the mouse's plantar surface, and when the mouse spontaneously lifted its hind paw, the pressure was recorded. Both left and right legs were stimulated alternately at 30-second intervals, and the measurements were repeated five times to calculate the hind paw retraction threshold.

In this experiment, chronic fibromyalgia was induced through intermittent cold stimulation, requiring strict control of temperature to prevent skin and muscle damage caused by low temperatures. It is important to note that this repeated chronic stimulation may induce anxiety, depression, and other emotions.

### *7.2.2 Chronic primary cephalic-facial pain*

The global incidence of primary cephalic-facial pain is relatively high, affecting approximately 0.9–5.1% of the population, with a tendency for recurrent attacks. The pathogenesis of this type of pain is complex [34]. Several methods have been developed to create animal models of chronic migraine, which we briefly outline in the sections that follow [35].

In one method, Melo-Carrillo et al. anesthetized rats and drilled into the right frontal bone. They fixed a bone wax sleeve directly above the skull, taking care to avoid damaging the meninges before closing the wound. After 2 days of recovery, 2 μL of an "inflammatory soup" (containing beta-imidazolylethylamine, 5-HOT, bradykinin1mmol/L, PGE 21 mmol/L) was injected. Behavioral changes were recorded before and after the injection. The results showed that the rats exhibited decreased exploratory behavior and increased nociceptive behavior, which could be alleviated by nonsteroidal anti-inflammatory drugs (NSAIDs) and the migraine treatment drug zolmitriptan.

Another popular method among researchers involves injecting drugs and migraine pathogenic genes [36] into mice. For example, Pradhan et al. [37] injected nitroglycerin (a mixture of 5 mg/mL glyceryl trinitrate, 30% ethanol, 30% propylene glycol, and saline to 10 mg/kg) intraperitoneally into mice for 9 days. Mechanical and thermal pain thresholds were measured before the injection, 75 minutes after the injection, and 2 hours after the injection. The results showed that the mice had reduced mechanical and thermal pain thresholds, slower weight gain, photophobia, and reduced activity. This mold-making method can be widely replicated and facilitates the observation of behavioral responses in rats in the awake state. However, the disadvantage is that it cannot specifically target intracranial vessels, and the systemic response is relatively large [38].

#### *7.2.3 Chronic primary visceral pain*

Abdominal pain is a common symptom in conditions like irritable bowel syndrome (IBS), often accompanied by changes in bowel habits [39]. It is currently believed that increased visceral sensitivity is the main pathophysiological characteristic of irritable bowel syndrome, which is manifested by a reduced threshold and increased sensitivity to stimuli [40]. To model this in animals, [41] some researchers have used the saccule expansion method to induce chronic visceral pain sensitivity in young rats. The procedure involves using a saccule tip to stimulate the anus of the young rat, ensuring complete defecation to prevent fecal mass interference. The saccule is inserted about 20 mm from the anus to the colon descendens, inflating the intestinal tract to maintain a pressure of 60 mmHg for 60 seconds. After gradually releasing the gas, the saccule is removed. The process is repeated every half hour, twice daily for 14 days. Rats are weaned at 4 weeks in groups of 4–6 per cage until the eighth week. Starting in the ninth week, colorectal dilation is performed to evaluate the abdominal wall withdrawal reflex, with higher scores indicating successful model induction. In addition, acupuncture at the "Shangjuxu (S37)" acupoint was found to alleviate chronic visceral pain sensitivity, potentially via regulation of the NGF/PI3K/TRPV1 signaling pathway.

Two other models used to study chronic visceral pain include colorectal distension in newborn rats and maternal-neonatal separation [42]. In the latter, newborn rats are separated from their mothers for 3 hours daily between postnatal days 2 and 21, with an ambient temperature maintained at 32 ± 0.5°C. After 22 days, the mice are weaned and housed in separate cages [43]. Both methods, focusing on mechanical stimulation and psychological stress, successfully induce chronic visceral pain [43] in adult rats.

#### *7.2.4 Chronic primary skeletal muscle pain*

Chronic primary skeletal muscle pain includes chronic primary neck, chest, waist, and limb pain [44]. Haiyan et al. [45] outlined several common animal molds for inducing cervical spondylotic radiculopathy, including pure compression models, joint stimulus models, and noninvasive intervention models.

Pure compressions models employ various techniques such as vascular clamping, silk thread ligation, compression with L-shaped stainless steel rods, silicon film compression, or autologous bone compression to compress the nerve roots.

Combined stimulation models combine compression and chemical stimulation to simulate the inflammatory response that occurs following nerve root compression in cervical spondylosis.

Noninvasive intervention models more accurately replicate the physiological curvature changes and stress imbalances in the cervical spine. In one method, researchers placed rabbits on a specialized fixation frame that maintained the cervical spine at 45° lower flexion using a head-flexion fixation method. The rabbits were kept in this position for 5 hours per day and then allowed to roam freely in their home cages. After 2 months, the model was established [46, 47]. Although this method minimizes animal injury, the experimental period is relatively long.

For lumbar pain models, common techniques include discogenic intervertebral body models, joint process joint-origin models, and nerve root injury models [48]. Kim [49] et al. created a lumbar pain model by damaging the L4-L5 and L5-L6 intervertebral discs of rats and removing nucleus pulposus tissue. Imaging and histological findings showed significant disc destruction and inflammatory cell infiltration.

Behavioral measures of pain showed a significant decrease in stress tolerance. Other models involve injecting substances into the joint cavity to damage cartilage [50], or surgically altering physiological structures to induce pathological changes [51].

### 7.2.5 Animal model of chronic secondary pain

Chronic secondary pain is often associated with underlying conditions such as cancer, postoperative or posttraumatic injuries, and neuropathic disorders [27].

### 7.2.6 Chronic cancer pain

Many patients with mid- to advanced-stage cancer experience persistent and progressively worsening pain, which seriously affects their physiological, psychological, and immune functions. Among the most common types of chronic cancer pain is osteosarcoma-related pain [52, 53].

In one study, researchers used a rat model to simulate cancer-induced bone pain. The procedure involved anesthetizing rats with an intraperitoneal injection of 10% chloral hydrate at a dose of 350 mg/kg. Then, 3 μl of breast cancer cells (approximately $4.8 \times 10^6$ cells/mL) were injected into the bone marrow cavity of the tibia, approximately 1 cm from the knee joint. The needle was withdrawn 5 minutes after the injection, and the entry point was quickly sealed with bone wax before suturing the layers. To verify the successful establishment of the cancer pain model, researchers observed the rats' spontaneous behaviors, assessed their thermal pain thresholds, and performed X-ray imaging and HE (hematoxylin-eosin) staining of the affected tibia on days 8 and 14.

There are many methods for inducing chronic cancer pain through tumor cell injections, but attention should be paid to the dosage and precise injection site to avoid complications.

### 7.2.7 Chronic postoperative or posttraumatic pain

Postoperative pain can severely impede a patient's recovery, with 10–50% of patients transitioning from acute pain to chronic pain, severely impacting their mental health and quality of life [54]. In one study, [55] researchers created a rat model to simulate postoperative pain. The procedure involved anesthetizing the rat and making a 1-cm vertical incision along the toe, about 0.5 cm from the right heel. The skin and subcutaneous fascia were cut, exposing the underlying muscles, which were then separated and lifted with tweezers. After ensuring hemostasis by pressing the area with a cotton swab, the skin was closed. To prevent infection, iodine-volt disinfection was applied, followed by erythromycin ointment.

Although creating a postoperative pain model is relatively straightforward, care must be taken to avoid local infection post surgery.

### 7.2.8 Chronic neuropathic pain

Mechanical injury is a common cause of neuropathic pain [56]. Meng et al. established a model of chronic neuropathic pain by ligating the spinal nerves. After anesthetizing the rats and placing them in the prone position, the researchers made an incision along the dorsal side after shaving the fur. The tissue was peeled back, and blunt dissection was performed to expose the spinal nerves. The fifth spinal nerve was

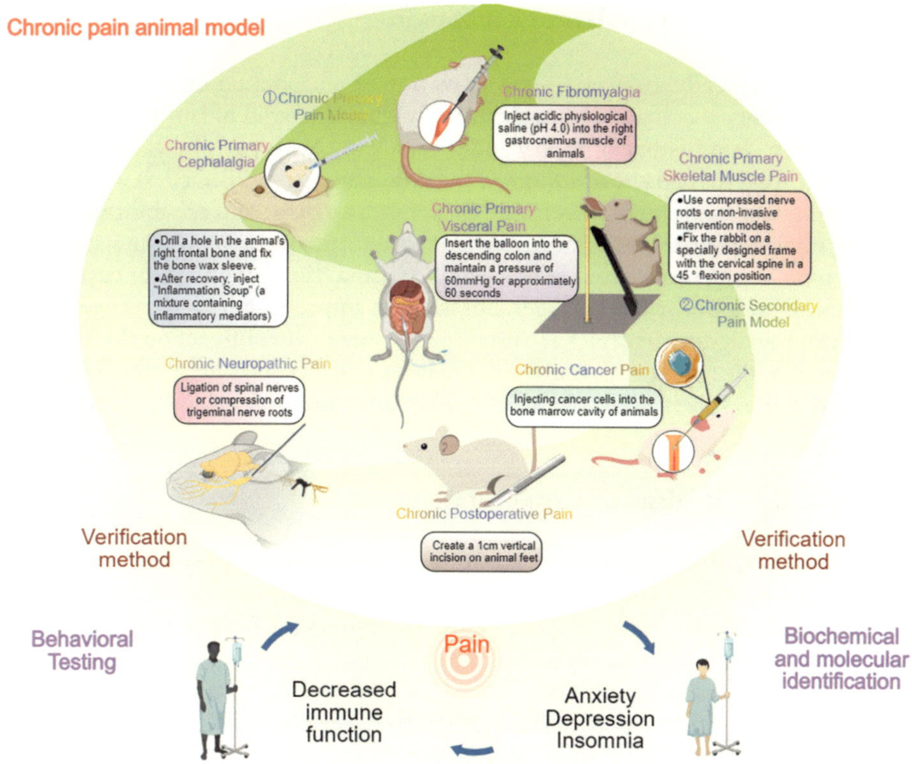


**Figure 1.**
*This schematic illustrates experimental models for studying chronic pain, including primary and secondary pain conditions in laboratory animals. The diagram demonstrates various pain induction methods, accompanied by two verification approaches: behavioral testing and biochemical/molecular identification. The outer circle depicts the interconnection between chronic pain, immune function, and psychological manifestations (anxiety, depression, and insomnia).*

ligated using a #5 suture, and the skin was sutured and sterilized. This model requires high precision to avoid damaging the nerve's activity.

Another animal model [57] focused on trigeminal neuralgia, where the right infraorbital nerve of rats was exposed. A thin plastic wire (0.1 cm in diameter) was inserted intracranially to compress the trigeminal nerve root. Pain behaviors in the rats were recorded for 4 weeks. However, this model has some limitations. It can only simulate chronic mechanical compression of the trigeminal nerve and does not replicate the complex microvascular physiological processes involved in human trigeminal neuralgia (**Figure 1**).

### 7.3 Summary and outlook

The creation of animal models of chronic pain is essential for studying the pathogenesis, diagnosis, and treatment of chronic pain-related diseases. However, some types of clinical pain cannot yet be fully replicated in animal models [16]. Therefore, further exploration of experimental design methods is necessary. A deeper integration of these methods with clinical practices will help foster the bidirectional development of both basic and clinical research, ultimately enhancing our understanding and treatment of chronic pain.

## 8. Advances in animal models of acute pain

Acute pain, typically induced by the activation of peripheral nociceptors, serves as an adaptive sensory mechanism that alerts individuals to avoid harmful stimuli. Acute pain often has a clear history of injury and disease [58], with peripheral mediators varying according to the type of injury. Clinically, acute pain is linked to various diseases, and its management is a critical aspect in treatment and rehabilitation. Targeting injury-specific mediators could improve future acute pain management.

Acute pain is often associated with surgical trauma, tissue damage, or certain disease states. Its appropriate management is key to improving patient recovery. While various analgesic therapies are used to minimize pain's impact on the body and internal organs, these treatments often come with side effects. Therefore, animal models play a vital role in studying pain mechanisms and evaluating the efficacy of analgesic drugs [16].

### 8.1 Associated mechanisms of acute orofacial pain

Prolonged peripheral nerve block has proven useful in managing both acute and chronic pain [59]. In studies of craniofacial-origin pain, various models have been used, including the rat acute pain model and the rat orofacial formalin test. These models often involve the eye-wiping test, a key indicator of trigeminal pain response. Researchers have tested local anesthetics (e.g., lidocaine, bupivacaine, and mepivacaine), general anesthetics (e.g., ketamine), muscle relaxants (e.g., baclofen), capsaicin, anti-inflammatory drugs (e.g., diclofenac), salicylates, antidepressants (e.g., amitriptyline and imipramine), and α2 adrenergic agents (e.g., clonidine). Combinations of these drugs have been explored for treating regional neuralgia. Local therapies tend to show fewer systemic side effects, minimal drug-drug interactions, and satisfactory efficacy [60].

As with the hot plate test [61], the eye-wiping experiment is an effective and appropriate model of acute pain in the trigeminal nervous system, It provides a rapid method of confirming evidence of analgesics in small cohorts, particularly for opioids and tricyclic antidepressants. This test has proven reliable for studying trigeminal neuralgia [59]. In addition, the orofacial formalin test is a valid and reliable model of nociception. One study compared two dosage forms of lidocaine, namely standard lidocaine and HP-β-cyclodextrin lidocaine, to assess the antinociceptive effects in different models of orofacial pain [62].

### 8.2 Mechanisms associated with acute gout arthralgia

The clinical presentation of gout is primarily due to the interaction between monosodium urate (MSU) crystals and local tissue, leading to an acute inflammatory response. Histological examinations of the synovium during a gout flare reveal hyperproliferation of the synovial layer and infiltration by neutrophils, mononuclear phagocytes, and lymphocytes. Acute gout attacks are frequently precipitated by specific factors, including trauma, surgery, concurrent illnesses, excessive alcohol intake, or medications that influence serum uric acid levels [63–65].

For the first time in mammals, it has been shown that injecting MSU crystals into the rat knee enhances joint nociceptor activity, paralleling behavioral signs of inflammation and pain similar to those observed in human gout attacks. This may serve as a useful animal model for studying gouty arthritis [64].

Pain is a hallmark of osteoarthritis, a complex "total joint" disease caused by inflammatory mediators rather than a simple "wear and tear" process. In addition to cartilage degeneration, the pathology of osteoarthritis involves synovitis, subchondral bone remodeling, ligament and meniscus degeneration, and joint capsule hypertrophy [66]. Using the experimental gouty arthritis model, researchers also explored the enhanced joint nociceptor activity associated with osteoarthritis [64].

To relieve gout symptoms, we synthesized WN1703, a xanthine oxidoreductase (XOR) inhibitor. In a rat model, WN1703 and febuxostat were shown to decrease inflammatory factors and reduce ankle swelling [10]. This study further investigated the anti-inflammatory effects of WN1703 in MSU-stimulated THP-1 cells and in rats with acute gout [67]. The results indicate that WN1703 reduced inflammation by inhibiting key proteins involved in the NLRP3/ASC/caspase-1 and TLR4/MyD88/NF-κB signaling pathways in both in vitro and in vivo models. Therefore, the anti-inflammatory activity of WN1703 may help in the treatment of gouty arthritis.

#### *8.2.1 Research progress in gout*

Gout is a common clinical disease characterized by hyperuricemia as the primary biochemical abnormality and the pathological deposition of MSU crystals in joints. This leads to high morbidity and disability rates, seriously affecting human health and causing acute pain. Therefore, understanding the pathogenesis of gout has become a growing area of research. Due to the many limitations of human trials, it is necessary to establish and study animal models of gout.

At present, common animal models used in gout include hyperuricemia models and gouty arthritis models. Each model has its unique characteristics and can reflect certain aspects of gout pathology. However, these models still show some limitations when compared to the full clinical presentation of gout (**Figure 2**) [68].

### 8.3 Mechanisms associated with acute cancer pain

Pain management in cancer patients, particularly postoperative acute pain, is challenging. Effectively managing this type of pain is crucial for improving patient outcomes. A comprehensive understanding of patient comorbidities, prior chronic pain experiences, and current analgesic usage is essential for devising effective strategies to mitigate the impact of acute pain [69]. Appropriate postoperative analgesia can reduce the risk of persistent postoperative pain and minimize long-term opioid dependence [70].

Experimental cancer models in mice have been developed to investigate the underlying mechanisms of acute cancer pain, highlighting a crucial interaction between protease-activated receptor 2 (PAR 2) and serine proteases. Serine proteases, such as trypsin, have been shown to trigger acute cancer pain through a PAR 2-dependent pathway. The progression of chronic cancer pain is associated with increased levels of serine proteases in the tumor microenvironment and the upregulation of PAR 2 in peripheral nerves. Suppressing serine protease activity has been found to alleviate persistent cancer pain severity and halt the development of chronic pain in PAR 2-deficient mice. Thus, targeting the interaction between PAR 2 and serine proteases represents a promising therapeutic approach for managing acute cancer pain and preventing its transition into chronic pain [69].

Carcinoma bone pain is an important factor affecting the quality of life of cancer survivors [71]. In women undergoing breast cancer surgery, acute postoperative pain can lead to adverse short-term and long-term outcomes [72]. Despite the rapid

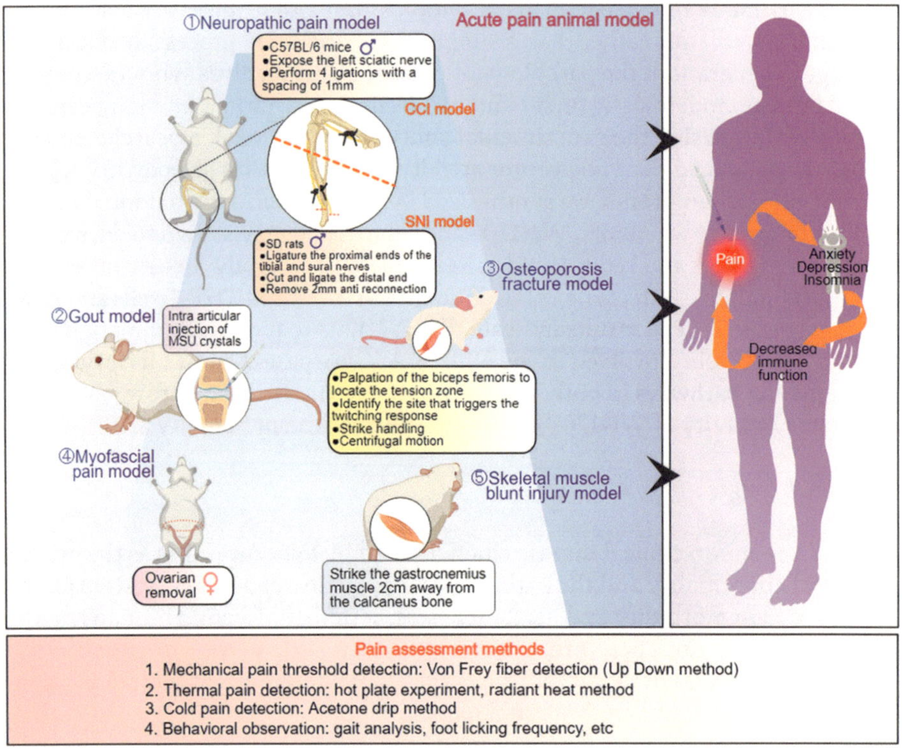


**Figure 2.**
*This figure illustrates experimental acute pain animal models and their behavioral consequences. The left panel demonstrates various rodent pain modeling approaches, from neuropathic to mechanical injury models. The right panel shows the resultant systemic effects, including anxiety, depression, and immune dysfunction. The bottom panel outlines standardized pain assessment methods, including mechanical, thermal, and behavioral evaluations. This system enables systematic investigation of pain mechanisms under controlled conditions.*

development of laparoscopic surgery techniques and perioperative management, managing postoperative pain remains an important clinical issue, particularly in patients undergoing laparoscopic colorectal cancer surgery [73]. Thus, the development of animal models to study acute postoperative pain in cancer patients is essential for informing clinical treatment strategies.

### *8.3.1 Other cancer pain models*

Certain cancers, such as colorectal cancer (CRC), can be influenced by dietary choices, suggesting potential avenues for prevention [74]. Animal models are instrumental in exploring clinical cancer prevention, especially for hepatocellular carcinoma (HCC), which is a fatal disease with limited therapeutic options in advanced stages. The long incubation period of chronic liver disease leading to HCC spans decades, presenting an important window for therapeutic opportunities to prevent HCC and improve patient outcomes. However, progress in developing clinical strategies for HCC chemoprevention has been slow, highlighting the value of animal models in studying tumorigenesis and development. These models provide insights into the molecular and genetic mechanisms involved in HCC progression [75].

The application of animal models extends to various forms of acute pain, such as orofacial neuralgia, gout, bone joint pain, cancer, and postoperative pain. These models have significant research value for determining analgesic drug choices, understanding the mechanisms of drug action, exploring targeted therapies, and developing strategies to prevent advanced cancer.

### 8.4 Summary and outlook

The use of reliable animal models is important in the study of acute pain mechanisms. However, there are many clinical pain conditions that cannot be accurately replicated through behavioral changes in animal models alone. To address these limitations, an increasing number of researchers are employing functional magnetic resonance imaging (fMRI) as a supplementary method to enhance the understanding of pain mechanisms that may not be fully captured by behavioral observations.

Despite this advancement, capturing the brain's response mechanisms to certain acute pain types remains challenging. For example, cardiac pain and acute angina are difficult to study using fMRI due to their brief onset and rapid relief, often occurring within a few minutes. Thus, there are limited studies investigating the central brain's response mechanisms to these conditions. This gap in understanding poses a new challenge for future studies on acute pain in animal models.

## 9. Application of animal models in the study of pain mechanisms

According to the International Association for the Study of Pain, pain is an unpleasant sensory and emotional experience associated with actual or potential tissue damage [1]. Compared to the old definition of pain, which emphasized the subjective nature of pain and tissue damage as the basis for pain, the modern version of the definition emphasizes the importance of the sensory nervous system for pain and focuses more on the individual's self-perception. The subjective nature of pain makes its measurement and assessment complex. A thorough knowledge of the sensory system's normal anatomy and physiology is crucial for understanding pain pathways.

### 9.1 Study on pain conduction mechanisms

1. Afferent sensory nerves relay different types of sensory information to the brain. Sensory receptors in the skin and tissues produce electrical impulses when activated. These signals are then transmitted to neural cell bodies in the dorsal root ganglion of the spinal cord and conveyed to the brain via the spinothalamic and spinobrachial pathways. Key brain regions involved include the parabrachial medulla, thalamus, amygdala, limbic system, and the somatosensory cortex. Pain signaling begins at the free nerve endings of primary afferent neurons, which are activated by various stimuli [76, 77].

2. The perception of pain involves four key processes: transduction, transmission, modulation, and perception. Transduction takes place in peripheral nerve endings, where primary afferent neurons are activated by harmful stimuli.

Primary afferent neurons possess receptors such as vanilloid receptor 1 (TRPV1), which detects heat, capsaicin, and protons, and Mas-related G protein-coupled receptors (Mrgprs), which are thought to mediate pain responses to mechanical stimuli. Pain signals are transmitted through two types of fibers: the fast-conducting, myelinated Aδ fibers, responsible for sharp, acute pain, and the slower, unmyelinated C fibers, which convey dull, lingering pain. Both fibers terminate in the dorsal horn of the spinal cord.

Aδ fibers connect with neurons in laminae I and V of the dorsal horn, while C fibers synapse with neurons in laminae I and II. The high degree of cellular plasticity in the dorsal horn enables modulation or "gating" of pain signals, which can occur in peripheral tissues, within the spinal cord, or in the brain, adjusting the pain response. The lateral spinothalamic tract transmits signals to the ventral posterolateral nucleus of the thalamus, relaying information about the duration, location, and intensity of pain.

In contrast, the medial spinothalamic tract carries pain signals to the medial thalamus, which is associated with autonomic and emotional aspects of pain perception. Pain perception is mediated by third-order neurons in the thalamus that project to cortical areas involved in both sensory and emotional aspects of pain. Modulation along these pathways can either enhance or inhibit pain perception.

3. Neuropathic pain is defined as "pain caused by impairment of the somatosensory system or disease" [78, 79], with prevalence estimates ranging from 7 to 9.8%, and up to 20% of patients experiencing chronic pain. Neuropathic pain results from [80, 81] spontaneous activity or abnormal responses to normal stimuli within a dysfunctional somatosensory system, often occurring without obvious tissue damage. Unfortunately, neuropathic may not respond well to pharmacological and noninterventional treatments, leading to severe disability and diminished quality of life.

There are many factors that contribute to neuropathic pain, including impairments in afferent pathways and changes in ion channel density and function, central and peripheral sensitization, cortical reorganization, and disinhibition of cellular and molecular changes. Sun et al. [82] used a neuropathic pain model in mice, combined with calcium signal detection, electrophysiology, optogenetics, and behavioral assessment, to demonstrate that the activity of LBPN glutamate neurons is crucial for transmitting both neuropathic and physiological pain. GABAergic neurons, which account for 10% of LPBN neurons, play a key role in gating the occurrence and transmission of neuralgia. This has important implications for a deeper understanding of neuralgia pathogenesis.

Moreover, the sympathetic nervous system appears to play a role in maintaining neuropathic pain [83]. This pain often manifests as abnormal pain, characterized by central pain sensitization [84] due to repetitive nonpainful stimuli. Peripheral sensitization involves the formation of ectopic foci, while central changes include reduced non-nociceptive input, downregulation of dorsal horn opioid substances and GABA_A receptors, and loss of layer II dorsal horn interneurons, leading to "disinhibition" of pain stimuli. Changes in central regulation can lead to hyperexcitability of damaged nerves, loss of C fibers, and increased activity of the sympathetic nervous system.

Effective treatments are designed to target receptors and neurotransmitters involved in these mechanisms. First-line drugs for neuropathic pain include tricyclic antidepressants, gabapentin (best for post-thermal neuralgia and diabetic neuropathy), and pregabalin (shorter time to treatment and lower effective doses).

### 9.2 Study on pain regulation mechanisms

1. The modulation of pain is an endogenous process thought to provide a survival advantage. This concept was described by anesthesiologist John Beecher during World War II, who observed that soldiers who had suffered severe war injuries often reported little or no pain. This suggests that the body has an endogenous mechanism that separates and regulates (enhances or decreases) pain transmission. The mechanisms for this phenomenon include segmental inhibition, the endogenous opioid system, and descending inhibition of the nervous system. In addition, cognitive and coping strategies also play a role in modifying pain perception.

More widely known and originally described by Melzack and Wall is "gated" segmental inhibition, which suggests that synapses between afferent neurons (Aδ and C fibers) and dorsal horn neurons in the spinal cord can be blocked when large myelinated nerve fibers (Aβ) that perceive touch (nonharmful stimuli) stimulate inhibitory nerves in the spinal cord. This inhibition reduces the transmission of pain signals by inhibiting the afferents of small unmyelinated C-fiber afferents. This explains why rubbing a wound reduces the feeling of pain. The mechanism by which transcutaneous electrical nerve stimulation is used for pain control is based on this theory [85].

2. Research has shown that pain is closely related to the body's stress reflex functions. Baroreceptors located in the carotid artery sinus and aortic arch can sense arterial blood pressure changes. These signals are transmitted to the central nervous system, which adjusts heart function and blood pressure to maintain cardiovascular homeostasis [86]. Patients with chronic hypertension tend to have a significantly higher pain tolerance compared to people without hypertension [87, 88]. This phenomenon suggests an interaction between the central nervous system pathways of blood pressure regulation and those that regulate pain.

The correlation between short-term blood pressure regulation and pain indicates a stress reflex's role in pain modulation. Xuan et al. proposed that interventions targeting baroreceptive reflex function in patients with acute and chronic pain could activate pain regulation pathways, potentially improving clinical symptoms [89].

Mood is also an essential factor in pain perception; increased depression can increase pain, leading to a vicious cycle [90]. Zhang MM professor [91] established a novel mouse model of chronic pain empathy, revealing that the enhanced glutamatergic projection pathway from the intercalated nucleus to the basolateral amygdala (IC-BLA) plays a significant role in the production and regulation of pain empathy. This study identified key synaptic signaling molecules involved in this process, offering new insights into the neural mechanisms underlying pain empathy.

Professor Wang Changhe's [92] long-term research on the interaction between pain and emotion revealed that the anterior cingulate cortex (ACC) and the ventral tegmental area (VTA) form a positive feedback loop that not only mediates the interaction between pain and emotion but also plays a critical role in the progression and maintenance of chronic pain. This suggests that targeting the ACC-VTA-ACC feedback loop may provide new therapeutic avenues for chronic pain management.

## 10. Application of animal models in the study of pain treatment strategies

### 10.1 Research on drug treatment strategies

Animal models play a vital role in early drug development, aiding in the understanding of disease mechanisms and the identification of potential therapeutic targets. Neuropathic pain, as defined by the International Association for the Study of Pain [78, 93], arises from injury or dysfunction of the somatosensory system and is characterized by a variety of underlying mechanisms and diverse clinical symptoms. Despite significant research, the development of drugs specifically targeting pain has been limited, with many therapies being repurposed from other indications. For example, opioids, anticonvulsants, and tricyclic antidepressants have demonstrated partial effectiveness in managing neuropathic pain, even though they were not initially developed for this purpose. Pregabalin, while showing modest benefits in specific neuropathic pain conditions, does not have strong clinical evidence supporting its use for generalized neuropathic pain. Lower-tier evidence suggests that a 600-mg dose of pregabalin may provide moderate pain relief for specific conditions, such as painful diabetic neuropathy (number needed to treat [NNT] = 6.3, 95% CI 4.6 to 10), postherpetic neuralgia (NNT = 5.6, 95% CI 3.5 to 10), and central neuropathic pain (NNT = 4.0, 95% CI 3.1 to 5.5). However, these figures indicate that only a minority of patients achieve meaningful pain relief with pregabalin, even for these particular disorders. In conclusion, while animal models are indispensable in early drug research, translating preclinical success into clinical effectiveness poses significant challenges. The case of pregabalin underscores the necessity for more targeted and rational drug development strategies, especially for conditions like neuropathic pain, which are driven by intricate and multifaceted pathophysiological mechanisms.

Traditional Chinese Medicine (TCM) provides distinctive approaches for pain management. Animal models of chronic pain have been instrumental in elucidating the pharmacological properties and analgesic mechanisms of TCM compounds and monomers [94, 95]. For instance, the analgesic effects of aconite soup may be attributed to its modulation of the MAPK signaling pathway, which reduces neuroinflammation. Additionally, *Eucommia ulmoides* polysaccharide has been shown to alleviate chronic constriction injury (CCI)-induced hyperalgesia by inhibiting glial fibrillary acidic protein (GFAP) activation and downregulating chemokine receptor 1 expression [96, 97]. Nonpharmacological treatments like acupuncture and massage are also widely utilized for pain management. Acupuncture, in particular, is well-regarded for its ability to regulate the autonomic nervous system. Stimulation of specific acupoints can alter autonomic nerve function, leading to pain relief [98, 99]. Studies using heart rate variability and fMRI have demonstrated that stimulation of the caudal ventrolateral medulla (CVLM) can induce changes in cardiovascular autonomic control [100, 101]. Similarly, traditional massage therapy has been shown to alleviate various types of pain, offering a nonpharmacological alternative for pain relief [102].

## 11. Challenges of animal pain modeling in pain research

### 11.1 Differences between animal and human pain

1. Animal models of nociception (pain) dating back to the late nineteenth century are critical to our understanding of the pain process. Since their introduction, various animal models have been developed to study pain across a range of diseases, covering both acute and chronic conditions. These models have been instrumental in advancing our understanding of disease-specific mechanisms and processes [103, 104]. However, a significant debate persists over which animal models are most appropriate for pain research and the optimal behavioral measures to use. This debate is rooted in the difficulty of translating preclinical findings into effective clinical therapies. For example, an editorial in *Nature Reviews Drug Discovery* (2012) reported a sharp decrease in the success rate of new compounds entering the market from phase 1 trials, falling from 10% between 2002 and 2004 to just 5% between 2006 and 2008 [105]. Similarly, a survey of clinical success rates for new drugs from 2003 to 2011 showed that nearly 90% of compounds were found unsuitable for human use before reaching phase 1 trials [106]. Despite these challenges, animal model research has led to notable breakthroughs, such as the development of tumor necrosis factor-α antibodies for rheumatoid arthritis and the use of N-type calcium channel blockers (ziconotide) and nerve growth factor inhibitors (tanezumab) for chronic pain management [107, 108]. Thus animal models should reflect clinical disease manifestations and pathology. For example, osteoarthritis knee pain usually includes mild pain at rest (spontaneous pain) but significant pain during exercise (evoked pain). In contrast, people with neuropathic pain usually have significant spontaneous pain and pain induced by touch or stress. Therefore, multiple outcome measures should be examined in animal studies to ensure the results mirror human pain conditions accurately [109].

2. Animal models of pain can serve as guides and complementary sources of information for human-based research. Animal models offer detailed insights into neurochemistry and anatomy with high temporal and spatial precision, allowing for direct electrophysiological recordings. Compared to human studies, they present distinct advantages, including better control over genetic and environmental variables, as well as improved safety and cost efficiency. Although DNA can be easily extracted from both animals and humans, obtaining mRNA from pain-specific tissues is usually only feasible in animals, except in rare clinical scenarios. Crucially, animal studies enable controlled investigations into chronic pain conditions that cannot be ethically or practically reproduced in human subjects. One distinctive aspect of certain animal models of peripheral neuropathic pain is partial denervation. While loosely ligating a peripheral nerve may not fully replicate human neuropathic pain, it remains one of the limited methods available to mimic this condition. Utilizing actual patients instead of healthy individuals may introduce confounding variables, making the interpretation of experimental data more challenging. A shared challenge in both human and animal pain research is that pain remains a subjective experience. Just as it is impossible to know what pain feels like in a mouse, we cannot fully grasp another person's pain experience. In both instances, pain is assessed based

on observed behaviors. However, humans have the unique capability to articulate their pain through introspective self-reporting, offering more direct insight into subjective experiences. By contrast, in animals, data collection is limited to observing behavioral responses, lacking the direct subjective input available in humans. Another advantage of using human subjects is their ability to remain still during imaging procedures without the need for training or anesthesia, unlike rodents. Human participants can also follow complex instructions, improving the reliability of experimental results compared to animal models. Animal models serve as a valuable complement to human neuroimaging, which has inherent technical limitations in terms of specificity and sensitivity [110]. Studies indicate that neuroimaging has limitations in detecting individual neuronal events and interpreting blood oxygen level-dependent signals, making it challenging to precisely examine small CNS regions. Therefore, animal models of pain should serve as guides and complementary sources of information for human-based research [18].

## 11.2 The difference between experimental and clinical pain

Despite the value of classical animal pain models, significant developmental changes have occurred in recent years [19]. First, there has been excessive emphasis on reflex withdrawal obtained from mechanical or thermal stimuli as dependent measures [111]. This measure is considered suboptimal due to its poor alignment with human symptoms and considerable experimental bias [112, 113]. Second, the focus has been predominantly on measuring pain itself, overlooking the important states of comorbid pain, including sequelae or comorbidity. Patients with chronic pain experience a range of issues, such as disability, anxiety, depression, cognitive dysfunction, insufficient sleep, loss of libido, and social withdrawal, which can complicate assessments in animals experiencing a chronic inflammatory or neuropathic state [19].

In recent years, there has been growing focus on comorbidities and associated symptoms, particularly in studies that include more comprehensive methodological details [114]. One notable criticism is that many existing models appear overly artificial, relying on agents such as formalin, carrageenan, and Freund's adjuvant for arthritis, or surgical nerve injuries to simulate neuropathic pain. However, progress has been made with models such as postoperative incisional pain [115] and cancer pain [116], which address these criticisms and improve the clinical relevance of preclinical research. These models align more closely with human pathological conditions. Additionally, a significant mismatch exists between the epidemiological reality of chronic pain prevalence in the human population and the usual selection of animal models. Most chronic pain patients are female [117], with a higher prevalence in middle-aged and older adults than in younger populations; however, pain studies predominantly use young adult male mice and rats due to convenience [19]. Furthermore, it has been suggested that design issues and reporting criteria in animal experiments fall short of those prevalent in human clinical trials. Specifically, details about blinding, randomization, and data loss are rarely reported in animal pain studies, potentially leading to high experimental bias. Indeed, only a subset of rodent pain studies is specifically designed to evaluate analgesic efficacy, yet those providing detailed methodological information significantly enhance transparency and facilitate replication efforts. An increasing number of journals now offer online methods sections or supplementary materials that are not constrained by traditional word limits, promoting comprehensive reporting [18].

### 11.3 Differability in pathways and group mechanisms in pain models

Past studies have often created different independent pain models for pain, but there is a growing awareness that chronic pain does not represent a single disease; rather, it comprises multiple pain causes [118]. Consequently, therapies targeting a single type of pain often do not yield expected results. A notable example is the neurokinin-1 (NK1) receptor antagonist (MK-869), along with other cases of glycine-site antagonists and sodium channel blockers for treating neuropathic pain, as well as neuronal gap junction blockers for migraine treatment. Recent studies emphasize the importance of understanding the group mechanisms involving multiple targets and pathways.

For example, in the study by Zhuo et al. [91] on the neural mechanisms underlying pain empathy, it was found that the excitability of neurons in the anterior cingulate cortex (ACC) and insular cortex (IC) increased during the early stages of pain empathy in littermate mice. However, only the IC neurons continued to enhance their activity during the later phases. These findings suggest that the ACC is primarily involved in the formation and induction of pain empathy, while the IC plays a crucial role in both the formation and consolidation phases.

Fu [119] et al. found that electrical stimulation of Zusanli acupoints activates a group of sensory neurons expressing the Prokr2 protein. Sensory neurons expressing Prokr2, stimulated at different intensities, activate different neural pathways. The low-intensity stimulation can activate the vagus-adrenal pathway, drive the adrenal gland to release catecholamine anti-inflammatory substances, inhibit the release of pro-inflammatory cytokines, and then significantly improve the survival rate of animals. High-intensity electroacupuncture stimuli will activate another group of sympathetic reflex, showing the bidirectional effect of the same type of neurons. This provides a new approach for managing inflammatory pain.

In clinical practice, acupuncture also plays a role in multi-target and multi-pathway regulation [120]. Huo et al. [121] found that stimulating the "Zusanli" point in rats activates nerve fibers, including Aβ and Aδ fibers, to produce analgesic effects. Furthermore, another study [122] pointed out that acupuncture can regulate the activity of microglia and astrocytes in the spinal cord's dorsal horn, thereby influencing neuron-glial communication and producing analgesia. Additionally, research [123] indicated that electroacupuncture increases the expression of the endocannabinoid receptor (CB1) in γ-aminobutyric acid (GABA) neurons in the brainstem, promoting serotonin release from the descending inhibitory pathway (5-HT) and reducing chronic pain. Therefore, we believe that the clustering effect of multiple pathways and targets will provide more methods and ideas for the treatment of pain.

## 12. Outlook for future research

Animal pain models are crucial to contemporary medical research, aiding in the understanding of human pain mechanisms and serving as a foundation for developing new pain therapies. As science and technology advance, future research on animal pain models will increasingly focus on precision, interdisciplinary collaboration, and the effectiveness of clinical translation.

## 12.1 More accurate animal models

A primary objective for future research in animal pain models is to develop more accurate and representative models. Animal models have been indispensable in biomedical research, providing crucial insights into disease mechanisms and serving as platforms to evaluate treatments. However, they also have inherent limitations. Pain research poses unique difficulties, primarily because pain is a subjective experience that is inherently challenging to replicate in animal models. While techniques such as the formalin and hot plate tests have been established to model pain, they often fall short of capturing the full complexity of human pain experiences. A key challenge lies in interspecies variability, which can lead to inconsistencies when translating findings from basic research to clinical applications. Variations in physiology, genetics, and metabolism between animals and humans can cause divergent treatment responses, underscoring the importance of precise model selection and rigorous data interpretation. Additionally, the complexity of human diseases is seldom mirrored accurately in animal models, which often oversimplify disease processes, thereby reducing the relevance of research outcomes. Despite notable advancements in pain research, substantial challenges persist in translating basic research findings into clinical practice. Animal pain models often do not fully capture the diversity of clinical pain types, including their causes and subjective sensations. Even when animal models approximate human pain, there is frequently a lack of precise tools to evaluate the various dimensions of the pain experience. Consequently, pain mechanisms identified in animal models may not translate effectively into clinical settings [124]. Furthermore, different animal models simulate specific aspects of human diseases and are suited for distinct research scenarios. The value of research findings hinges significantly on whether the chosen model aligns with the specific research objectives. There is no universal model to simulate all types of pain, thus researchers must select models that are best suited to their particular study requirements. Conventional pain models frequently fail to capture the complexity of human pain, prompting researchers to create more sophisticated and diverse models. For example, gene-editing technologies such as CRISPR/Cas9 can precisely alter animal genes to generate models that more closely replicate human pain conditions [125]. Additionally, advanced imaging techniques, such as functional MRI (fMRI) and electrophysiological methods, enable researchers to directly observe pain signal transmission and processing in animals, facilitating the development of more refined models.

## 12.2 Multidisciplinary intersection

Cross-disciplinary cooperation will greatly promote the development of animal pain models. The deep integration of the field of pain research with neuroscience, molecular biology, computer science, and other disciplines will bring new perspectives and methods to the development of pain models. By combining artificial intelligence and big data analysis, researchers can accurately characterize pain signal transmission pathways and intelligently analyze pain behavior data, leading to the prediction and screening of more effective pain treatment strategies.

## 12.3 Summary of discussion

The ultimate goal of research in animal pain models is to achieve more precise clinical translation. Findings from animal models must be effectively translated into clinical practice to address actual pain issues faced by patients. The advancement of basic medicine should be closely aligned with clinical practice to ensure that laboratory results can be validated and applied in clinical settings. Furthermore, basic research findings should inform clinical practice to uncover new problems, thereby optimizing the reciprocal development of both fields in order to better serve experimental research and patient treatment.

For example, conducting preclinical trials allows for comprehensive assessments of the efficacy and safety of drugs or treatments before formal human application. Additionally, retrospective studies of drugs already in use can provide valuable insights. Future research on animal pain models is expected to make significant breakthroughs in precision, interdisciplinary collaboration, and clinical transformation. This may include direct interventions in relevant brain regions using brain-computer interfaces to alleviate corresponding issues, as well as targeted modulation of specific brain areas using technologies such as high-intensity focused ultrasound (HIFU). Such advancements will accelerate the research and development of new drugs and technologies, ultimately providing more effective treatments for pain patients.

## Author details

Zhichao Wang, Lumei Xiang, Xiaoxia Wang, Xiaoqing Tan and Anfeng Xiang*
Affiliated Hospital of Traditional Chinese Medicine, Three Gorges University, China

*Address all correspondence to: afxiang0101@163.com

## References


[1] Raja SN et al. The revised International Association for the Study of Pain definition of pain: Concepts, challenges, and compromises. Pain. 2020;**161**(9):1976-1982

[2] Martucci KT, Mackey SC. Neuroimaging of pain: Human evidence and clinical relevance of central nervous system processes and modulation. Anesthesiology. 2018;**128**(6):1241-1254

[3] Edwards RR et al. The role of psychosocial processes in the development and maintenance of chronic pain. The Journal of Pain. 2016;**17**(9 Suppl.):T70-T92

[4] Lin X, Qiu D, Xu J. Discussion on experimental animal models of pain. China Practical Medicine. 2007;**34**:146-149

[5] Barrett JE. The pain of pain: Challenges of animal behavior models. European Journal of Pharmacology. 2015;**753**:183-190

[6] Percie DSN, Rice AS. Improving the translation of analgesic drugs to the clinic: Animal models of neuropathic pain. British Journal of Pharmacology. 2014;**171**(12):2951-2963

[7] McGonigle P, Ruggeri B. Animal models of human disease: Challenges in enabling translation. Biochemical Pharmacology. 2014;**87**(1):162-171

[8] Davis KD, Moayedi M. Central mechanisms of pain revealed through functional and structural MRI. Journal of Neuroimmune Pharmacology. 2013;**8**(3):518-534

[9] Seath CP et al. Tracking chromatin state changes using nanoscale photo-proximity labelling. Nature. 2023;**616**(7957):574-580

[10] Zou Y. Experimental Zoology. Science Press; 2012

[11] Care NRCU, Animals AUOL. Guide for the care and use of laboratory animals. In: The National Academies Collection: Reports Funded by National Institutes of Health. Washington (DC): National Academies Press (US); 2011

[12] Brown, Marilyn J, et al. Culture of care: Organizational responsibilities. Management of Animal Care and Use Programs in Research, Education, and Testing. 2017:11-26

[13] National Certification and Accreditation Standardization Technical Committee Laboratory Accreditation Subcommittee, etc., GB/T 27416-2014. Understanding and Implementing the General Requirements for the Quality and Capacity of Laboratory Animal Institutions. China: Standards Press; 2015

[14] Li Hong, Fu Qiang, Huang Qingzhen, et al. Microbiology, parasitology and quality control of experimental animals [J]. Medical Animal Control, 2008;(07):527-528

[15] Huang D, Feng P. Selection of experimental animals in biomedical research. Journal of Anhui Education University (Natural Science Edition). 1999;**01**:37-38

[16] Liu X, Gong L. Progress in animal models based on acute and chronic pain. Modern Doctor of China. 2023;**61**(27):126-129

[17] Zhang M, Luo Y. Overview of animal models of pain-related diseases. Western Medicine. 2014;**26**(06):814-817

[18] Mogil JS, Davis KD, Derbyshire SW. The necessity of animal models in pain research. Pain. 2010;**151**(1):12-17

[19] Mogil JS. Animal models of pain: Progress and challenges. Nature Reviews. Neuroscience. 2009;**10**(4):283-294

[20] Zhuo M. Cortical plasticity as a new endpoint measurement for chronic pain. Molecular Pain. 2011;**7**:54

[21] Jackson T et al. A systematic review and meta-analysis of the global burden of chronic pain without clear Etiology in low- and middle-income countries: Trends in heterogeneous data and a proposal for new assessment methods. Anesthesia and Analgesia. 2016;**123**(3):739-748

[22] Kawai K et al. Adverse impacts of chronic pain on health-related quality of life, work productivity, depression and anxiety in a community-based study. Family Practice. 2017;**34**(6):656-661

[23] Dai Q et al. Progress in animal models of chronic pain and their application in traditional Chinese medicine research. Chinese Journal of Traditional Chinese Medicine. 2020;**45**(24):5866-5876

[24] Chen Y, Chen J. Progress in animal models of inflammatory pain. Chinese Journal of Pharmacology and Toxicology. 2023;**37**(S1):35

[25] Boscán P et al. A dog model to study ovary, ovarian ligament and visceral pain. Veterinary Anaesthesia and Analgesia. 2011;**38**(3):260-266

[26] Avona A et al. Dural calcitonin gene-related peptide produces female-specific responses in rodent migraine models. The Journal of Neuroscience. 2019;**39**(22):4323-4331

[27] Treede RD et al. Chronic pain as a symptom or a disease: The IASP classification of chronic pain for the international classification of diseases (ICD-11). Pain. 2019;**160**(1):19-27

[28] Nicholas M et al. The IASP classification of chronic pain for ICD-11: Chronic primary pain. Pain. 2019;**160**(1):28-37

[29] Compagnoni R et al. Fibromyalgia and shoulder surgery: A systematic review and a critical appraisal of the literature. Journal of Clinical Medicine. 2019;**8**(10):1518-1518

[30] Sarzi-Puttini P et al. Fibromyalgia: An update on clinical characteristics, aetiopathogenesis and treatment. Nature Reviews Rheumatology. 2020;**16**(11):645-660

[31] Zhu Q. Consensus of Chinese experts on clinical diagnosis and treatment of fibromyalgia. Chinese Journal of Pain Medicine. 2021;**27**(10):721-727

[32] Yen CM, Hsieh CL, Lin YW. Electroacupuncture reduces chronic fibromyalgia pain through attenuation of transient receptor potential vanilloid 1 signaling pathway in mouse brains. Iranian Journal of Basic Medical Sciences. 2020;**23**(7):894-900

[33] Wakatsuki K et al. Repeated cold stress, an animal model for fibromyalgia, elicits proprioceptor-induced chronic pain with microglial activation in mice. Journal of Neuroinflammation. 2024;**21**(1):25

[34] Qin G, Chen L, Zhou J. Research progress on chronic migraine. Chinese Journal of Pain Medicine. 2011;**17**(03):176-178

[35] Melo-Carrillo A, Lopez-Avila A. A chronic animal model of migraine,

induced by repeated meningeal nociception, characterized by a behavioral and pharmacological approach. Cephalalgia. 2013;**33**(13):1096-1105

[36] van den Maagdenberg AM et al. A Cacna1a knockin migraine mouse model with increased susceptibility to cortical spreading depression. Neuron. 2004;**41**(5):701-710

[37] Pradhan AA et al. Characterization of a novel model of chronic migraine. Pain. 2014;**155**(2):269-274

[38] Ren L et al. Progress in the research of animal experimental models of chronic migraine. China Clinical Neuroscience. 2018;**26**(03):332-335

[39] Camilleri M. Diagnosis and treatment of irritable bowel syndrome: A review. JAMA. 2021;**325**(9):865-877

[40] Liangbing W, Jiarong G, Ting W, et al. Study on the therapeutic effect of ethanol extract on regulating the CRF-CREB signaling pathway on visceral hypersensitive rats with irritable bowel syndrome [J]. Pharmacology and Clinical Practice of Traditional Chinese Medicine, 2017;**33**(06):96-101

[41] Chen C et al. NGF/PI3K/TRPV1 pathway mediates acupuncture "Shangjuxu" to regulate visceral pain in rats with irritable bowel syndrome chronic visceral pain sensitivity model. Acupuncture Research. 2023;**48**(10):1017-1024

[42] Ren TH et al. Effects of neonatal maternal separation on neurochemical and sensory response to colonic distension in a rat model of irritable bowel syndrome. American Journal of Physiology. Gastrointestinal and Liver Physiology. 2007;**292**(3):G849-G856

[43] Tang Y, Lin C, Chen Y. Comparison of the modeling effects of two chronic visceral pain models. Journal of Fujian Medical University. 2007;**06**:582-585

[44] Lu Y et al. Chronic primary pain. Chinese Journal of Pain Medicine. 2021;**27**(02):81-86

[45] Cheng H et al. Progress in research on animal models of cervical radiculopathy. Chinese Journal of Laboratory Zoology. 2023;**31**(01):91-97

[46] Hou L, Zhang R, Wang Y. Neck and shoulder massage reduces pain and depression levels in employees with type II cervical spondylosis. Basic Medicine and Clinic. 2018;**38**(08):1141-1144

[47] Ying H, Chen L, Zhan H, Jing F, Lu R. Establishment of an experimental non-invasive rabbit cervical disc degeneration model. Chinese bone. Trauma. 2004;**08**

[48] Li H et al. Research progress on animal models related to low back pain. Chinese Journal of Laboratory Zoology. 2019;**27**(03):399-404

[49] Kim JS et al. The rat intervertebral disk degeneration pain model: Relationships between biological and structural alterations and pain. Arthritis Research & Therapy. 2011;**13**(5):R165-R165

[50] Gong K et al. Rat model of lumbar facet joint osteoarthritis associated with facet-mediated mechanical hyperalgesia induced by intra-articular injection of monosodium iodoacetate. Journal of the Formosan Medical Association. 2011;**110**(3):145-152

[51] Zhang JJ et al. Autologous nucleus pulposus transplantation to lumbar 5 dorsal root ganglion after epineurium dissection in rats: A modified model

of non-compressive lumbar herniated intervertebral disc. Chinese Medical Journal. 2011;**124**(13):2009-2014

[52] Singh V, Gillespie TW, Harvey RD. Intranasal ketamine and its potential role in cancer-related pain. Pharmacotherapy. 2018;**38**(3):390-401

[53] Junxia Zhang et al. Investigated the mechanism of low-dose ketamine inhibiting morphine tolerance in rats with cancer pain. Journal of Inner Mongolia Medical University. 2022;**44**(01):61-65

[54] Wang J et al. Electroacupuncture alleviates hyperalgesia by regulating CB1 receptor of spinal cord in incisional neck pain rats. Evidence-based Complementary and Alternative Medicine. 2021;**2021**:5880690

[55] Lu Z et al. Effect of electroacupuncture pretreatment on 5-HT_7 receptor expression in midbrain periaqueductal gray in rats with incision pain. Chinese Journal of Pain Medicine. 2024;**30**(02):94-99

[56] Binder A, Baron R. The pharmacological therapy of chronic neuropathic pain. Deutsches Ärzteblatt International. 2016;**113**(37):616-625

[57] Zhou R. An animal model of trigeminal neuralgia was established by chronic compression of the trigeminal nerve root. Chinese Journal of Pain Medicine. 2012;**18**(12):766

[58] Regmi B, Shah MK. Possible implications of animal models for the assessment of visceral pain. Animal Models and Experimental Medicine. 2020;**3**(3):215-228

[59] Farazifard R et al. Eye-wiping test: A sensitive animal model for acute trigeminal pain studies. Brain Research. Brain Research Protocols. 2005;**16**(1-3):44-49

[60] Xu J, Brennan TJ. The pathophysiology of acute pain: Animal models. Current Opinion in Anaesthesiology. 2011;**24**(5):508-514

[61] Inaltekin A, Kivrak Y. Evaluation of the effect of Vortioxetine on pain threshold by hot-plate test in mice. Noro Psikiyatri Arsivi. 2021;**58**(4):274-277

[62] de Oliveira SB et al. Comparison of antinociceptive effects of plain lidocaine versus lidocaine complexed with hydroxypropyl-beta-cyclodextrin in animal models of acute and persistent orofacial pain. Naunyn-Schmiedeberg's Archives of Pharmacology. 2019;**392**(5):573-583

[63] Dalbeth N, Haskard DO. Mechanisms of inflammation in gout. Rheumatology (Oxford). 2005;**44**(9):1090-1096

[64] Marcotti A et al. Joint nociceptor nerve activity and pain in an animal model of acute gout and its modulation by intra-articular hyaluronan. Pain. 2018;**159**(4):739-748

[65] Liu F et al. Cryo-shocked tumor cells deliver CRISPR-Cas9 for lung cancer regression by synthetic lethality. Science Advances. 2024;**10**(13):eadk8264

[66] Ayhan E, Kesmezacar H, Akgun I. Intraarticular injections (corticosteroid, hyaluronic acid, platelet rich plasma) for the knee osteoarthritis. World Journal of Orthopedics. 2014;**5**(3):351-361

[67] Yasin ZNM. Macrophage polarization in THP-1 cell line and primary monocytes: A systematic review. Differentiation. 2022;**128**:67-82

[68] Jian G, Qing Y, Zhang Q. Research status and progress on animal

models of gout. Medical Review. 2021;**27**(12):2397-2401

[69] Cata JP et al. Postoperative acute pain challenges in patients with cancer. Best Practice & Research. Clinical Anaesthesiology. 2019;**33**(3):361-371

[70] Lam DK et al. Novel animal models of acute and chronic cancer pain: A pivotal role for PAR2. The Journal of Neuroscience. 2012;**32**(41):14178-14183

[71] Zheng XQ et al. Neurophysiological mechanisms of cancer-induced bone pain. Journal of Advanced Research. 2022;**35**:117-127

[72] Raza MM et al. Chronic breast pain prior to breast cancer surgery is associated with worse acute postoperative pain outcomes. Journal of Clinical Medicine. 2021;**10**(9):1887-1887

[73] Oh EJ et al. Analgesic efficacy of Nefopam as an adjuvant in patient-controlled analgesia for acute postoperative pain after laparoscopic colorectal cancer surgery. Journal of Clinical Medicine. 2021;**10**(2):270-270

[74] Terasaki M et al. Fucoxanthin and colorectal cancer prevention. Cancers (Basel). 2021;**13**:10

[75] Shankaraiah RC et al. Animal models of hepatocellular carcinoma prevention. Cancers (Basel). 2019;**11**(11):1792-1792

[76] Sneddon LU. Comparative physiology of nociception and pain. Physiology (Bethesda). 2018;**33**(1):63-73

[77] Yam MF et al. General pathways of pain sensation and the major neurotransmitters involved in pain regulation. International Journal of Molecular Sciences. 2018;**19**(8):2164-2164

[78] Treede RD et al. Neuropathic pain: Redefinition and a grading system for clinical and research purposes. Neurology. 2008;**70**(18):1630-1635

[79] Finnerup NB et al. Neuropathic pain: An updated grading system for research and clinical practice. Pain. 2016;**157**(8):1599-1606

[80] Bouhassira D et al. Prevalence of chronic pain with neuropathic characteristics in the general population. Pain. 2008;**136**(3):380-387

[81] Yawn BP et al. The prevalence of neuropathic pain: Clinical evaluation compared with screening tools in a community population. Pain Medicine. 2009;**10**(3):586-593

[82] Sun L et al. Parabrachial nucleus circuit governs neuropathic pain-like behavior. Nature Communications. 2020;**11**(1):5974

[83] Finnerup NB, Jensen TS. Mechanisms of disease: Mechanism-based classification of neuropathic pain-a critical analysis. Nature Clinical Practice. Neurology. 2006;**2**(2):107-115

[84] Nickel FT et al. Mechanisms of neuropathic pain. European Neuropsychopharmacology. 2012;**22**(2):81-91

[85] Lee GI, Neumeister MW. Pain: Pathways and physiology. Clinics in Plastic Surgery. 2020;**47**(2):173-180

[86] Rabinovitch A et al. The Baroreflex mechanism revisited. Bulletin of Mathematical Biology. 2015;**77**(8):1521-1538

[87] Ghione S et al. Arterial hypertension is associated with hypalgesia in humans. Hypertension. 1988;**12**(5):491-497

[88] Nascimento RM et al. Pressure pain threshold is higher in hypertensive compared with normotensive older adults: A case-control study. Geriatrics & Gerontology International. 2017;**17**(6):967-972

[89] Cui Q, Che L, Xu L. Mechanism and clinical application of baroreflex in pain management. Union Medical Journal. 2024;**15**(02):258-264

[90] Uckun AC et al. The role of pain catastrophizing and depression in the outcomes of physical therapy in a prospective osteoarthritis cohort. Pain Physician. 2020;**23**(2):209-218

[91] Zhang MM et al. Glutamatergic synapses from the insular cortex to the basolateral amygdala encode observational pain. Neuron. 2022;**110**(12):1993-2008.e6

[92] Song Q et al. An ACC-VTA-ACC positive-feedback loop mediates the persistence of neuropathic pain and emotional consequences. Nature Neuroscience. 2024;**27**(2):272-285

[93] Jensen TS et al. A new definition of neuropathic pain. Pain. 2011;**152**(10):2204-2205

[94] Phillips TE, Cherry CL, Cox S, et al. Pharmacological treatment of painful HIV-associated sensory neuropathy: A systematic review and meta-analysis of randomised controlled trials. PLoS One. 2010;**5**(12):e14433

[95] Guo Q et al. Study on the regulatory mechanism of inflammatory network of Wuzi decoction in alleviating neuropathic pain. Chinese Journal of Pharmacy. 2019;**54**(06):1054-1061

[96] Liu Y et al. Effect of Huangqi Guizhi Wuwu decoction on the expression of CX3CR1 in spinal cord in rats with neuropathic pain. Modern Journal of Integrated Traditional Chinese and Western Medicine. 2020;**29**(07):723-727

[97] Jiang Z, Zhou M. Analgesic effect of euonymus polysaccharide on neuropathic pain model rats. Journal of Gannan Medical College. 2019;**39**(02):114-117

[98] Li YW et al. The autonomic nervous system: A potential link to the efficacy of acupuncture. Frontiers in Neuroscience. 2022;**16**:1038945

[99] Pang Y et al. Regulated aberrant amygdala functional connectivity in premenstrual syndrome via electro-acupuncture stimulation at sanyinjiao acupoint (SP6). Gynecological Endocrinology. 2021;**37**(4):315-319

[100] Dhond RP et al. Acupuncture modulates resting state connectivity in default and sensorimotor brain networks. Pain. 2008;**136**(3):407-418

[101] Beissner F et al. Acupuncture–deep pain with an autonomic dimension? NeuroImage. 2012;**60**(1):653-660

[102] Mak, Selene, et al. Use of massage therapy for pain, 2018-2023: A systematic review. JAMA Network Open. 2024;**7**(7):e2422259

[103] Berberich P, Hoheisel U, Mense S. Effects of a carrageenan-induced myositis on the discharge properties of group III and IV muscle receptors in the cat. Journal of Neurophysiology. 1988;**59**(5):1395-1409

[104] Bennett GJ, Xie YK. A peripheral mononeuropathy in rat that produces disorders of pain sensation like those seen in man. Pain. 1988;**33**(1):87-107

[105] A decade in drug discovery. Nature Reviews. Drug Discovery. 2012;**11**(1):3

[106] Roelcke V. The animal model of human disease as a core concept of medical research: Historical cases, failures, and some epistemological considerations. Science in Context. 2022;**35**(2):173-197

[107] Katz N et al. Efficacy and safety of tanezumab in the treatment of chronic low back pain. Pain. 2011;**152**(10):2248-2258

[108] Schroeder CI, Craik DJ. Therapeutic potential of conopeptides. Future Medicinal Chemistry. 2012;**4**(10):1243-1255

[109] Gregory NS et al. An overview of animal models of pain: Disease models and outcome measures. The Journal of Pain. 2013;**14**(11):1255-1269

[110] Davis KD. Neurophysiological and anatomical considerations in functional imaging of pain. Pain. 2003;**105**(1-2):1-3

[111] Vierck CJ, Hansson PT, Yezierski RP. Clinical and pre-clinical pain assessment: Are we measuring the same thing? Pain. 2008;**135**(1-2):7-10

[112] Chesler EJ et al. Identification and ranking of genetic and laboratory environment factors influencing a behavioral trait, thermal nociception, via computational analysis of a large data archive. Neuroscience and Biobehavioral Reviews. 2002;**26**(8):907-923

[113] Chesler EJ et al. Influences of laboratory environment on behavior. Nature Neuroscience. 2002;**5**(11):1101-1102

[114] He R et al. Research progress on animal models of pain and depression comorbidity and evaluation methods. Laboratory Animals and Comparative Medicine. 2022;**42**(01):68-73

[115] Brennan TJ, Vandermeulen EP, Gebhart GF. Characterization of a rat model of incisional pain. Pain. 1996;**64**(3):493-502

[116] Schwei MJ et al. Neurochemical and cellular reorganization of the spinal cord in a murine model of bone cancer pain. The Journal of Neuroscience. 1999;**19**(24):10886-10897

[117] Berkley KJ. Sex differences in pain. The Behavioral and Brain Sciences. 1997;**20**(3):371-380; discussion 435-513

[118] Abboud C et al. Animal models of pain: Diversity and benefits. Journal of Neuroscience Methods. 2021;**348**:108997

[119] Liu S et al. A neuroanatomical basis for electroacupuncture to drive the vagal-adrenal axis. Nature. 2021;**598**(7882):641-645

[120] Dai B et al. Combination of acupuncture therapy and exercise therapy-taking pain rehabilitation as an example. Acupuncture in China. 2024:876-880

[121] Huo R et al. Responses of primary afferent Fibers to acupuncture-like peripheral stimulation at different frequencies: Characterization by single-unit recording in rats. Neuroscience Bulletin. 2020;**36**(8):907-918

[122] Duanmu C et al. Progress in the research on acupuncture analgesia and the role of spinal glial cells in chronic pain. Chinese Journal of Basic Medicine of Traditional Chinese Medicine. 2017;**23**(03):443-446

[123] Yuan XC et al. Electroacupuncture potentiates cannabinoid receptor-mediated descending inhibitory control in a mouse model of knee osteoarthritis. Frontiers in Molecular Neuroscience. 2018;**11**:112

[124] Junchao Luo et al. Animal model for tendinopathy. Journal of Orthopaedic Translation. 2023;**42**:43-56

[125] Ailioaie LM, Litscher G. Molecular and cellular mechanisms of arthritis in children and adults: New perspectives on applied photobiomodulation. International Journal of Molecular Sciences. 2020;**21**(18):6565-6565

Chapter 5

# Exploring Animal Models for Interstitial Cystitis/Bladder Pain Syndrome

*Tadeja Kuret*

## Abstract

Interstitial cystitis/bladder pain syndrome (IC/BPS) is a chronic inflammatory disease of the urinary bladder characterized by discomfort and pain, increased urinary frequency, urgency, and nocturia. Most currently available treatment options primarily aim to alleviate clinical symptoms, with no single option providing a long-term beneficial effect for all patients. This limitation is likely due to the complex and multifactorial nature of IC/BPS and the incomplete understanding of its pathobiology. Preclinical studies using animal models remain essential for researching the etiology and pathophysiology of IC/BPS, discovering novel drug targets, and designing future clinical trials. Various animal models have been developed to replicate the primary symptoms and complex pathophysiology of IC/BPS, each with its own advantages and disadvantages. Given the disease's complexity and the existence of several distinct clinical subtypes, it is unlikely that a single model can fully replicate all aspects of IC/BPS. Instead, multiple distinct animal models will likely be necessary, depending on the subtype being evaluated. When using animal models to inform future clinical trials, special care must be given to understanding the specific underlying mechanisms of development and progression of a particular IC/BPS subtype and the mechanism of action of the therapeutic being studied. This approach will help ensure the successful translation of preclinical findings into clinical settings.

**Keywords:** interstitial cystitis/bladder pain syndrome, experimental/animal models, pathophysiology, therapy, translational potential

## 1. Introduction

Interstitial cystitis/bladder pain syndrome (IC/BPS) is a chronic inflammatory disease of the urinary bladder, without bacterial infection or a clear pathologic cause. Common symptoms include pain, increased urinary frequency, urgency, and nocturia [1]. The prevalence of IC/BPS is estimated to range from 10 to 300 per 100,000 individuals, though it is likely underreported due to the lack of standardized diagnostic criteria and reliable biomarkers [2, 3]. The condition is more prevalent in females than in males, with a reported ratio between 10:1 and 3:1 [4]. Patients with IC/BPS frequently have comorbid conditions such as autoimmune diseases, fibromyalgia, chronic fatigue syndrome, irritable bowel syndrome, and chronic psychological stress [5].

The etiology and pathophysiology of IC/BPS are still not well understood, with various hypotheses emerging, including bladder urothelium injury, increased barrier permeability, chronic inflammation with immune cell infiltration, oxidative stress, central sensitization, and possible autoimmune involvement [6]. The complexity of these mechanisms, along with the various different clinical presentations and the use of non-targeted therapies, hampers effective treatment. Although multiple therapeutic options exist, they only aim to relieve symptoms, and no single treatment has proven consistently effective for long-term management [7].

Given the unclear pathophysiology of IC/BPS, developing definitive therapies remains challenging. Consequently, substantial efforts have been directed toward creating experimental animal models that closely mimic the IC/BPS phenotype. To investigate pathophysiological mechanisms of IC/BPS, three main types of animal models have been developed: (1) bladder-centric models, (2) models with complex autoimmune mechanisms, and (3) models incorporating psychological and physical stressors or natural diseases. As no single model can encompass all aspects of the human condition, a combination of models is necessary to gain a comprehensive understanding of IC/BPS pathophysiology and progression, ultimately leading to the identification of effective treatment targets [8, 9].

This chapter provides an overview of IC/BPS, its diagnosis, classification, therapeutic management, and describes the pathophysiology of the disease. In the main part, commonly used experimental *in vivo* models for studying IC/BPS and evaluating therapeutic options are described. The chapter concludes by discussing the translational potential of these models and a look for the future.

## 2. Diagnosis and classification

The diagnosis of IC/BPS is currently based primarily on the exclusion of other diseases and conditions, with diagnostic practices varying significantly across different clinical centers worldwide [10]. According to the American Urological Association (AUA), the recommended approach for diagnosing IC/BPS includes a thorough patient history, physical examination, and laboratory tests to document symptoms and exclude other potential causes such as overactive bladder, undifferentiated chronic pelvic pain, endometriosis, and other comorbid chronic diseases. IC/BPS is diagnosed if a patient has experienced chronic pelvic pain, pressure, or discomfort associated with bladder filling for more than six months, accompanied by at least one additional urinary tract symptom, such as increased urinary frequency, urgency, or nocturia. When the diagnosis remains uncertain, cystoscopy and/or urodynamics may be performed [5].

IC/BPS can be broadly categorized into two major subtypes based on bladder histological findings: 1) Hunner type (ulcerative) IC/BPS (HIC/BPS) and 2) non-Hunner type (non-ulcerative) IC/BPS (NHIC/BPS). HIC/BPS is defined by the presence of Hunner's lesions, that is, mucosal lesions accompanied by abnormal capillary structures, within the bladder. This subtype is characterized by severe bladder-centric symptoms, reduced bladder capacity, urothelial denudation, inflammatory infiltrates, and edema. In contrast, NHIC/BPS lacks Hunner's lesions and shows minimal histological changes in the bladder. Patients with NHIC/BPS often experience systemic comorbidities and "bladder-beyond" pain, with no obvious bladder etiology. The classification of IC/BPS into these subtypes is determined by cystoscopy and histologic findings from bladder biopsies. HIC/BPS typically presents with more severe symptoms directly related to bladder pathology, whereas NHIC/BPS often presents with a broader spectrum of symptoms that may involve other systems. Understanding these subtypes is crucial for tailoring treatment approaches, as HIC/BPS may require therapies specifically targeting bladder inflammation and lesions, while NHIC/BPS might benefit from more systemic treatment strategies due to its association with wider comorbidities [11, 12].

## 3. Therapy

The AUA guidelines recommend a stepwise therapeutic approach, in which the first-line therapy includes patient education with daily behavior modification and lifestyle change including efficient stress management, changes in the diet, bladder training, and physical therapy. Second-line therapy involves oral administration of amitriptyline for pain management, cimetidine, antihistamines, and pentosan polysulfate sodium, which provides a protective coating to the bladder lining. Intravesical application of dimethyl sulfoxide (DMSO) that has anti-inflammatory and analgesic properties, heparin and lidocaine cocktails, which sooth the bladder lining, are also considered as a second-line therapy. Third-line therapy requires cystoscopy and hydrodistension, while neuromodulation and intravesical injection of botulinum toxin A (BTX-A) are considered as fourth- and fifth-line therapies. If a patient does not respond to any of the therapeutic agents, cyclosporine A is used as an immunosuppressive agent for refractory cases. The last resort is bladder augmentation by major surgical intervention [5].

The currently available and recommended therapy options for IC/BPS patients are primarily based on empirical studies and often have limited efficacy. As a result, ongoing research on IC/BPS is increasingly focused on the development and evaluation of novel therapeutic options, with experimental animal models playing a crucial role in these efforts.

## 4. Pathophysiology

The pathophysiology of IC/BPS is complex and multifactorial, and it remains incompletely understood. Several theories have been proposed to explain the disease's development and progression. It is still unclear whether IC/BPS arises primarily from an injury or disorder within the bladder, particularly within the urothelium, or if the bladder symptoms associated with IC/BPS develop secondary to another cause [5]. Various mechanistic studies have demonstrated that IC/BPS can result from local injury to the urothelial cell layer, leading to urothelial dysfunction, increased permeability that allows toxic substances from urine to penetrate the bladder wall, activation of the local bladder immune response, and chronic inflammation [13].

Conversely, IC/BPS is a common comorbidity with other chronic non-urologic inflammatory diseases, suggesting that immune system dysfunction and systemic inflammation might be common underlying causes for both IC/BPS and these comorbidities [14, 15]. Systemic inflammation has been shown to lead to peripheral afferent nerve hyperexcitability, causing symptoms such as urinary frequency and urgency, which suggests that systemic inflammation may be a primary cause of IC/BPS [16].

### 4.1 Urothelial cell damage and loss of function

One of the most common features observed in bladder biopsies from IC/BPS patients is the denudation or thinning of the bladder urothelium, including the complete or partial loss of the superficial cell layer [17]. In a healthy bladder, the urothelium functions as a tight permeability barrier, which is achieved by tight junction proteins, such as occludins and claudins, and specific transmembrane proteins uroplakins and a glycosaminoglycan (GAG) layer on the surface of superficial urothelial cells. By maintaining cell integrity, the urothelium ensures that the bladder remains impermeable to potentially harmful substances present in urine [18].

In patients with IC/BPS, a damaged urothelium with disrupted barrier function and increased permeability results from the reduced GAG layer and deregulated expression of tight junction proteins (zonula occludens-1 (ZO-1), occludin), adherens proteins (E-cadherin), and uroplakins [19–22]. Scanning electron microscopy has revealed severe defects in the integrity of the superficial urothelial cell layer, which correlate with increased disease severity and higher numbers of inflammatory cell infiltrates [23]. The increased urothelial permeability allows urine solutes, such as potassium and urea, to leak into the bladder wall, triggering an immune response, characterized by the release of proinflammatory cytokines such as interleukin (IL)1, IL6, and IL8 and tumor necrosis factor alpha (TNFα). These cytokines help to recruit and activate immune cells, particularly mast cells, in the bladder wall. In addition, urothelial cells also secrete a variety of signaling molecules, such as acetylcholine, adenosine triphosphate (ATP), and nitric oxide. In bladder biopsies from IC/PBS patients, increased nerve fiber number and density together with elevated levels of nerve growth factor (NGF) have been observed, pointing toward neurogenic inflammation and bladder afferent hypersensitivity [6, 24].

### 4.2 Chronic inflammation

Bladder biopsies obtained from IC/BPS patients have shown extensive infiltration of various immune cells, including macrophages, neutrophils, and mast cells. Notably, bladders of HIC/BPS patients have increased numbers of B- and T-lymphocytes and plasma cells [25]. Using single-cell RNA sequencing and spatial transcriptomics, Peng et al. identified 22 immune subpopulations in the bladders of IC/BPS patients. These included increased proportions of M2 macrophages, inflammatory CD14+ macrophages, conventional dendritic cells, central memory CD4+ T cells, activated B cells, and neutrophils. A substantial number of these immune cell populations were located within the urothelial cell region in IC/BPS bladders but not in control bladders, suggesting a key role of urothelial cells in the development of IC/BPS [26].

Various studies have shown that when urothelial cells are triggered *in vitro* by inflammatory stimuli, they increase the production and release of various proinflammatory mediators (e.g., IL6, IL8, CXCL1, CXCL10) and other molecules primarily involved in the innate immune response, such as Toll-like receptors (TLRs), NOD-like receptors, and NFκB [27–29]. The significant release of these proinflammatory mediators further damages urothelial cells, leading to increased permeability and a perpetuating cycle of inflammation and prolonged sensitization of peripheral afferent nerve endings within the bladder wall (**Figure 1**) [8].

Elevated urinary levels of several proinflammatory cytokines (e.g., IL1, IL6, IL8, TNFα) and chemokines (CXCL1, CXCL10, CCL2, CCL5, CCL11) have been measured in IC/BPS patients compared to controls [30, 31]. Additionally, higher serum/plasma

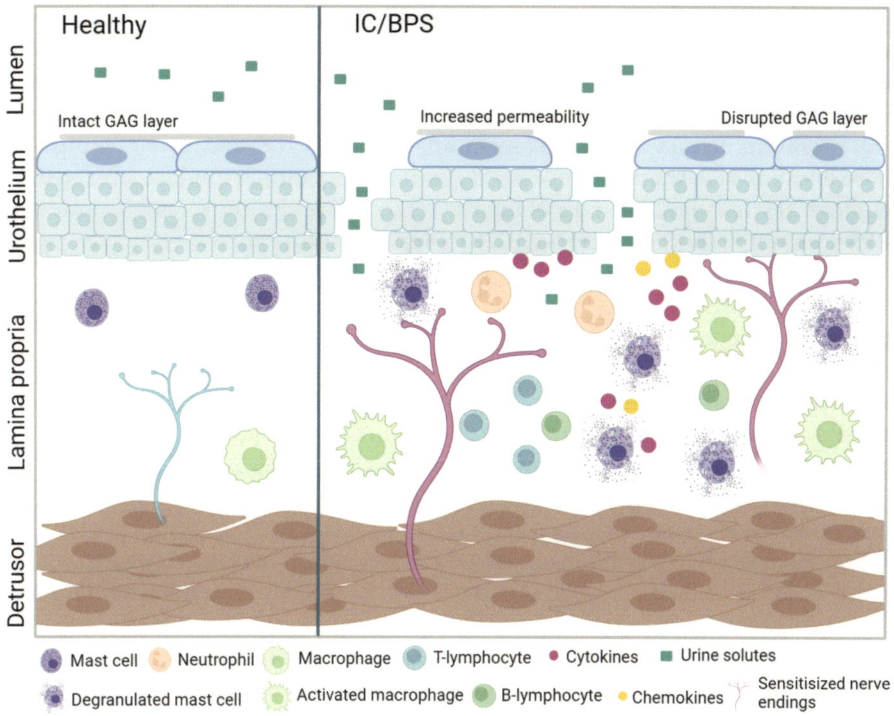


**Figure 1.**
*Schematic illustration of possible mechanisms of IC/BPS development. IC/BPS, interstitial cystitis/bladder pain syndrome; GAG, glycosaminoglycan. Figure was created using Biorender.com.*

levels of proinflammatory markers, such as C-reactive protein (CRP) and IL6, have been found in IC/BPS patients, indicating systemic inflammation [32, 33].

### 4.3 Autoimmune mechanisms

Indirect evidence, such as the high prevalence of IC/BPS in females and its clinical association with autoimmune diseases, supports the possibility of an autoimmune nature of the disease. IC/BPS patients have a significantly increased risk of developing Sjogren's syndrome, systemic lupus erythematosus, and rheumatoid arthritis [14].

Additionally, the presence of autoantibodies against bladder epithelial proteins has been detected in IC/BPS patients [34]. For example, anti-muscarinic receptor 3 autoantibodies, which are present in patients with Sjogren's syndrome, are proposed to bind to the M3 receptor on detrusor muscle cells in the bladder, potentially triggering an inflammatory response that contributes to IC/BPS development [35].

To further understand this connection, experimental animal models with complex autoimmune mechanisms have been developed to mimic the pathophysiology of IC/BPS [36].

### 4.4 Chronic psychological stress

Recent studies have highlighted the significant role of chronic psychological stress in the development and progression of IC/BPS. Patients with a history of early life trauma report greater sensory pain, higher levels of depression, and poorer physical quality of life [37]. IC/BPS patients are 2.7 times more likely to have experienced post-traumatic stress disorder (PTSD) compared to controls [38]. The link between IC/BPS, chronic stress, and inflammation is further supported by findings that IC/BPS patients exposed to both early and recent life trauma have increased levels of several proinflammatory mediators, including IL1β, IL6, IL8, and TNFα, compared to those without such traumatic events [39]. Moreover, animal models subjected to psychological stress have been developed to recapitulate chronic stress, resulting in an IC/BPS phenotype similar to the NHIC/BPS subtype [40].

## 5. Animal models

Although several theories have been proposed, the exact etiology and pathobiology of IC/BPS remain subjects of ongoing exploration. Over the past 20 years, various animal models have been developed to investigate the mechanisms underlying IC/BPS development and to explore potential therapeutic interventions. Most of these models utilize rodents, such as mice and rats, to simulate both acute and chronic forms of IC/BPS. In acute IC/BPS models, a single induction or inductions repeated for less than three days with bladder-toxic substances are used. Chronic experimental IC/BPS models involve treatments lasting more than three days or more complex mechanisms of induction. These animal models can be categorized into three groups: 1) bladder-centric models, which involve direct bladder injury or insult; 2) models with complex autoimmune mechanisms, mimicking dysfunctional immune system; and 3) models involving psychological and physical stressors or natural diseases, which simulate the impact of stress and comorbid conditions on IC/BPS development [8, 9, 40]. A summary of animal models used for studying IC/BPS is provided in **Table 1**.

### 5.1 Bladder-centric models

Bladder-centric models are the most common types of animal models used to create an IC/BPS phenotype. In these models, bladder inflammation with subsequent symptoms of pain, urinary urgency and frequency is induced via intravesical exposure to bladder irritants/toxic substances. The most commonly used substances are cyclophosphamide (CYP); chemical irritants, including hydrochloric acid (HCl); and protamine sulfate (PS) or bacterial products, for example, lipopolysaccharide (LPS).

#### *5.1.1 Cyclophosphamide (CYP)-induced*

CYP is a chemotherapeutic agent used in the treatment of various cancers, including lymphomas, leukemias, myeloma, lung cancer, and breast cancer, as well as some autoimmune diseases. In the liver, CYP is metabolized into acrolein, a toxic byproduct that is excreted in the urine. As acrolein accumulates in the bladder, it interacts with the urothelium, inducing an inflammatory response. Consequently, patients treated with CYP often develop chronic bladder inflammation and hemorrhagic cystitis [107].

Acute CYP treatment in rodents involves administering a single high dose (150–300 mg/kg) intraperitoneally [41–54]. Within 24 hours of CYP injection, significant bladder inflammation is observed, characterized by edema and infiltration of leukocytes and mast cells [42, 46, 49, 51, 53]. Additionally, CYP-treated animals show

| Reference | Animals | IC induction | Main findings |
|---|---|---|---|
| *CYP-induced acute models* | | | |
| Boudieu 2019 [41]<br>Zhang 2017 [42]<br>Liu 2020 [43]<br>Oliveira 2016 [44] | C57BL/6 J mice<br>males n = 32<br>females n = 56<br>females ND<br>females ND | 150 mg/kg,<br>i.p. 300 mg/kg, i.p. | edema, increased IL1β, IL6, IL8, TNFα, NFκB, MPO activity, COX2, oxidative stress, urothelial cell injury, increased number of voids & hyperactivity to von Frey filament |
| Chen 2014 [45]<br>Smaldone 2009 [46]<br>Chopra 2005 [47]<br>Liu 2015 [48]<br>Chen 2020 [49]<br>Coelho 2015 [50]<br>Auge 2013 [51] | SD rats<br>males n = 18<br>females ND<br>females ND<br>males ND<br>females n = 30<br>females n = 66<br>females n = 60 | 150 mg/kg, i.p. | edema, mast cells, increased IL1α/β, IL6, IL10, IL12, IL18, TNFα, GM-CSF, NFκB, MMP9, RANTES, iNOS, MIF, COX2, NGF, urothelial injury, erosion, ulceration, hemorrhage, reduced TEER & ZO1, permeability to water and urea, increased voiding frequency & non-voiding contractions, decreased ICI & voided volume, pain, mechanical hyperalgesia |
| Pessina 2015 [52]<br>Juszczak 2010 [53]<br>Malley 2002 [54] | Wistar rats<br>females n = 132<br>females n = 42<br>females n = 30 | 200 mg/kg,<br>i.p. 150 mg/kg, i.p. | edema, mast cells, increased IL1β, IL2, IL4, IL6, TNFα/β, mucosal abrasion, hemorrhage, increased number of voids, pain |
| *CYP-induced chronic models* | | | |
| deBerry 2015 [55]<br>Boudes 2011 [56]<br>Golubeva [57]<br>Peskar 2023 [58] | C57BL/6 mice<br>females ND<br>males n = 32<br>females n = 16<br>males & females<br>n = 20 | 100 mg/kg, i.p.,<br>1x/2 days, 5 days<br>40–80 mg/kg, i.p.,<br>1x/2 days, 7 days<br>80 mg/kg, i.p.,<br>1x/2 days, 7 days<br>1x/2 days, 8 days | mild inflammation, edema, mast cells, increased IL6, IL10, TNFα, MPO, CCL2, increased TNF-, TLR-, JAK/STAT-signaling, complement and coagulation cascades, cell cycle regulation, urothelial hyperplasia & thickness, increased urinary frequency, mechanical hyperalgesia |
| Auge 2021 [59]<br>Mahal 2018 [60]<br>Wang 2017 [61]<br>Li et al., 2020 [62]<br>Luo et al., 2020 [63]<br>Xie et al., 2018 [64]<br>Yang et al., 2021 [65]<br>Ko et al., 2021 [66] | SD rats<br>females n = 48<br>females n = 24<br>females n = 33<br>females n = 30<br>females n = 80<br>females n = 30<br>females ND<br>females n = 40 | 40–75 mg/kg, i.p.,<br>1x/3 days, 6 days<br>50 mg/kg, i.p., 2x/<br>week, 2 weeks<br>75 mg/kg, i.p.,<br>day 1 and 4<br>1x/3 days, 10 days<br>1x/3 days, 12 days<br>days 1, 3 and 5<br>1x/3 days, 7 days<br>1x/3 days, 9 days | inflammation, edema, mast cells, increased IL6, IL8, TNFα, NFκB, COX2, NGF, urothelial cell damage & disrupted integrity, increased urinary frequency & pressure, decreased ICI & cystometric capacity, hyperalgesia, visceral pain, nociceptive behavior |
| *HCl-induced models* | | | |
| Funahashi 2014 [67]<br>Song 2015 [68]<br>Kim 2017 [69]<br>Shimizu 2013 [70]<br>Konkol 2016 [71]<br>Hirose 2016 [72] | SD rats<br>females n = 18<br>females n = 45<br>females n = 45<br>females ND<br>females n = 52<br>F344 rats<br>females n = 60 | Acute<br>[0.1–0.4 M],<br>intravesically, | edema, PMN & mast cells, increased TNFα, IL1β, IL6, angiogenesis, tissue fibrosis, increased urothelial permeability & denudation, decreased UPK3A, increased urinary frequency & non-voiding contractions, decreased ICI & voided volume, nociceptive behavior |
| Furuta 2018 [73] | F344 rats<br>females n = 90 | Chronic (0.1 M],<br>intravesically<br>1x/week, 2 weeks | edema, mast cells, increased TNFα, MPO activity, submucosal hemorrhage, tissue fibrosis, increased TGFβ & collagen fibers decreased ICI, nociceptive behavior |

| Reference | Animals | IC induction | Main findings |
|---|---|---|---|
| *PS-induced models* | | | |
| Greenwood 2018 (75) | SD rats<br>females ND | 1 mg/ml,<br>intravesically | decreased permeability |
| Tyagi 2008 [74]<br>Tyagi 2009 [75]<br>Fraser 2003 [76] | SD rats<br>females ND<br>females ND<br>females n = 24<br>females n = 34 | 10 mg/ml,<br>intravesically, | increased bladder contraction frequency, decreased ICI |
| Soler 2008 [77] | Wistar rats<br>females n = 108 | 5%, intravesically, | edema, PMN infiltration |
| Centinel 2010 [78] | Wistar rats<br>females n = 32 | 10 mg/ml,<br>intravesically,<br>2x in 24 h | mast cells, increased MDA, decreased GSH, decreased permeability, irregular uroplakin distribution, dilated tight junctions |
| Akin 2015 [79] | SD rats<br>females ND | 5 mg/ml,<br>intravesically,<br>2x/day, 3 days | edema, lipid peroxidation, decreased GSH, CAT and SOD activity, increased urinary frequency, decreased voided volume |
| Grundy 2020 [80] | C57BL/6 J mice<br>females n = 27 | 1 mg/ml,<br>intravesically | increased bladder permeability<br>hypersensitivity of bladder afferents to bladder filling |
| *LPS-induced models* | | | |
| Gonzales 2005 [81]<br>Yoshizumi 2021 [82] | C57BL6 mice<br>females ND<br>SD rats<br>females ND | 0.1–1 mg/ml<br>intravesically | edema, PMN cells, increased TNFα, NGF, decreased ICI, pain |
| Tambaro 2014 [83] | CD-1 mice<br>males ND | 25 mg/kg, i.p. | leukocytes, increased IL1α/β, TNFα, MPO activity |
| Berger 2019 [84] | CD-1 mice<br>females n = 29<br>BALB/c mice<br>females n = 22 | 20 mg/kg,<br>i.p. 0.15 mg/ml,<br>intravesically | adherent leukocytes in bladder venules, inflammation, mechanical allodynia |
| Li 2017 [85]<br>Song 2019 [86] | SD rats<br>females n = 60<br>females n = 70 | 2 mg/ml,<br>intravesically<br>0.75 mg/ml,<br>intravesically | edema, mast cells, increased IL6, IL10, TNFα, IFNγ, NFκB, fibrosis, urothelial ulceration, hemorrhage, decreased micturition function & bladder capacity |
| Shih 2021 [87] | BALB/c mice<br>females n = 18 | 0.15 mg/ml,<br>intravesically, 2x/<br>week, 5 weeks | increased NLRP3, IL1β, TGFβ1, fibrosis, bladder injury, increased micturition |
| Ryu 2018 [88] | SD rats<br>females ND | 0.75 mg/ml,<br>intravesically, 1x/<br>week, 5 weeks | mast cells, urothelial denudation, increased non-voiding contractions |
| *EAC models* | | | |
| Liu 2019 [89]<br>Jin 2017 [83]<br>Singh 2013 [90]<br>Lin et al., 2008 [91] | C57BL/6 mice<br>females n = 30<br>females n = 40<br>SWXJ mice<br>females n = 20<br>females n = 20 | homogenized mice bladders, s.c. | increased bladder wall thickness, edema, leukocytes, neutrophils, mast cells, CD4+ T-lymphocytes,<br>increased IL1β, IL4, IL6, IL10, IFNγ, TNFα, CXCL9, CXCL10, CXCL11, increased urinary frequency, decreased ICI & voided volume, pain |

| Reference | Animals | IC induction | Main findings |
|---|---|---|---|
| Bicer 2015 [92]<br>Li 2019 [93]<br>Izgi 2013 [94]<br>Kim 2020 [95]<br>Chung 2019 [96]<br>Altuntas 2012 [97] | BALB/c mice<br>females n = 30<br>females n = 40<br>females n = 20<br>SD rats<br>females n = 24<br>females n = 25<br>SWXJ mice<br>females n = 10 | UP3A, s.c.<br>1 µg/µl<br>200 µg<br>200 µg<br>200 µg<br>UP2A, s.c.<br>200 µg | mast cells, T-lymphocytes, increased IL1β, IL6, IL17, TNFα, IFNγ, CCL2, MPO, NGF, TLR2, TLR4, TLR5, TLR11, autoantibodies, fibrosis, increased UP3, decreased ZO1, increased urinary frequency, decreased ICI & voided volume, mechanical hyperalgesia, pain |
| Liu 2008 [98]<br>Cui 2019 [99]<br>Kogan 2008 [100]<br>Liu 2007 [98]<br>Kim 2010 [101] | URO-OVA mice<br>females ND<br>females n = 30<br>URO-OVA/O-T<br>females n = 23<br>URO-OVA and URO-OVA/O-T<br>females ND<br>females ND | | edema, hyperplasia, mast cells, CD4+ and CD8+ T-lymphocytes, increased IL1β, IL6, TNFα, IFNγ, CCL2, NGF and substance P, increased urinary frequency, decreased voided volume, pelvic pain |
| *WAS-induced models* | | | |
| Zeybek 2007 [102]<br>Holschneider 2020 [103]<br>Lee 2015 [104]<br>Bazi 2012 [105] | Wistar Kyoto rats<br>females ND<br>females n = 20<br>females n = 22<br>SD rats<br>females n = 20 | 2 h/day, 5 days<br>1 h/day, 10 days<br>1 h/day, 10 days<br>2 h/day, 7 days | mast cells, increased MDA, decreased GSH, impaired urothelial integrity, increased urinary frequency, pain, hypersensitivity and stress-related cerebral activations within the supraspinal micturition circuit |
| West 2021 [106] | C57Bl/6 J mice<br>females n = 14 | 1 h/day, 10 days | increased number of voids, decrease in void size |

*CAT, catalase; CCL, CC chemokine ligand; COX2, cyclooxygenase-2; CXCL, C-X-C motif ligand; GM-CSF, granulocyte-macrophage colony-stimulating factor; GSH, glutathione; ICI, intercontraction interval; IFN, interferon; IL, interleukin; iNOS, inducible nitric oxide synthase; i.p., intraperitoneal; LPS, lipopolysaccharide; MDA, malondialdehyde; MIF, macrophage migration inhibitory factor; MMP, matrix metalloproteinase; MPO, myeloperoxidase; ND, not defined; NFκB, nuclear factor kappa B; NGF, nerve growth factor; NLRP3, NLR family pyrin domain containing 3; OVA, ovalbumin; PMN, polymorphonuclear; PS, protamine sulfate; s.c., subcutaneous; SOD, superoxide dismutase; TEER, transepithelial resistance; TGFβ, transforming growth factor beta; TNF, tumor necrosis factor; TLR, toll-like receptor; UP, uroplakin; ZO1, Zonnula occludens 1; WAS, water avoidance stress.*

**Table 1.**
*A summary of experiment animal models used for studying IC/BPS.*

significantly increased expression of proinflammatory mediators IL1β, IL6, TNFα, IL8, NFκB, and NGF in bladder tissues [41, 43, 46, 49–51, 54]. Markers of oxidative stress are also elevated in the bladders of these animals [43, 44]. The acute injection of a high CYP dose leads to substantial alterations within the bladder, such as damage and thickening of the urothelial cell layer, partial or complete loss of superficial urothelial cells, reduced transepithelial resistance, decreased expression of ZO1, and increased permeability to water and urea [47, 49]. One day after acute CYP injections, experimental animals exhibit increased urinary frequency with increased number of voids, decreased mean voided volume [44, 48, 49, 52], nociceptive behavior, hyperactivity to von Frey filaments, hyperalgesia, and pelvic pain [41, 43, 50–52]. Given the extent of bladder damage following a single high dose of CYP, which includes hemorrhages, urothelial cell destruction, and hematuria, it has been proposed that this model may be more relevant for studying acute hemorrhagic cystitis rather than IC/BPS [9].

To more accurately mimic chronic IC/BPS, an alternative model involving repeated systemic injections of lower doses of CYP (ranging from 40 mg/kg to 100 mg/kg) every 2 to 3 days over an extended period (typically 7 to 21 days) has been proposed [55–66]. This model induces milder chronic pain without significant adverse effects on body weight and overall animal health [9, 59]. It is characterized by a mild inflammatory response in bladder tissue, marked by edema of the lamina propria; moderate increases in the expression of proinflammatory markers IL1β, IL6, IL8, TNFα, and NGF; and mast cell infiltration [56, 57, 61–64, 66]. Bladder overactivity with decreased voiding function, increased voiding frequency, reduced voiding volume, mechanical hyperalgesia, and nociceptive behavior are also observed after repeated CYP administration [55, 56, 59, 60, 62, 64–66]. Importantly, no signs of tissue hemorrhages, hematuria, mucosal ulcerations, or urothelium loss are observed [56, 57, 59], suggesting that this model better represents the specific bladder changes seen in IC/BPS. A comprehensive transcriptome analysis of bladder tissue from a chronic model of CYP-treated mice revealed significant alterations in the transcriptome profile of urinary bladders. Enrichment analysis indicated an increase in the expression of genes involved in innate immunity pathways, including cytokine-cytokine receptor interactions, natural killer cell-mediated cytotoxicity, TNFα-, TLR, JAK/STAT-signaling, complement and coagulation cascades, and cell cycle regulation. These innate immunity pathways mirror those identified in transcriptome profiling of bladder samples from IC/BPS patients, underscoring the model's relevance for understanding IC/BPS mechanisms [58]. However, most studies show that the effects of CYP administration diminish approximately 10 days after treatment, indicating rapid tissue regeneration. As a result, even chronic CYP instillations cannot precisely mimic the persistence and progression of IC/BPS seen in human patients. Despite these limitations, the CYP-induced IC/BPS model remains one of the most commonly used and well-characterized bladder-centric models for studying IC/BPS [40].

#### *5.1.2 Chemical irritants-induced*

A 1–10-minute intravesical instillation of HCl (ranging from 0.1 to 0.4 M) results in urothelial denudation, with complete or partial loss of superficial urothelial cells and increased urothelial permeability as early as 1 to 3 days after exposure [69]. Funahashi et al. found decreased expression of uroplakin 3A and damaged tight junctions in the HCl group [67]. Bladder inflammation was evident with hemorrhage, edema, infiltration of immune cells (neutrophils, monocytes, lymphocytes and mast cells), and increased expression of proinflammatory IL1, IL6, and TNFα [68–73]. One to two weeks after the insult, bladder inflammation progressed, showing signs of angiogenesis and tissue fibrosis with increased expression of TGFβ and collagen fibers and thickening of the bladder wall [68, 69, 72, 73]. Regarding bladder function, animals with HCl-induced cystitis exhibited shorter intercontraction intervals, increased urinary frequency, decreased voided volume, irregular non-voiding bladder contractions, and nociceptive behavior [68–70, 72, 73]. These bladder changes persisted even two weeks after HCl instillation, recapitulating the disease progression observed in human IC/BPS patients.

PS is an arginine-rich protein used to counteract or reverse the anticoagulant effect of heparin. *In vivo* bladder instillation of PS (1 mg/ml to 10 mg/ml) is a controlled model of selective urothelial injury, which typically resolves within 5–10 days, and is useful for studying urothelial damage repair and tissue regeneration [108, 109]. PS acts by interrupting the GAG layer of urothelial cells, thus reducing transepithelial resistance and increasing urothelial cell permeability. It also disrupts urothelial tight junctions and affects the expression of uroplakins [78, 109]. Bladder inflammation following PS instillation is evidenced by edema and increased infiltration of polymorphonuclear and mast cells [77–79]. Additionally, PS induces oxidative stress in the bladder, as shown by elevated levels of malondialdehyde, a marker of lipid peroxidation, and decreased levels of antioxidants like glutathione, catalase, and superoxide dismutase [78, 79]. Animals with PS-induced cystitis exhibit increased urinary frequency, decreased voided volume and intercontraction intervals, and visceral pain [74–76, 79]. Grundy et al. recently demonstrated that bladder afferent hypersensitivity to bladder filling occurs one day after PS stimuli, even without underlying bladder inflammation. However, mechanosensitive responses to distension returned to normal levels seven days after PS treatment, indicating rapid urothelial regeneration [80].

#### *5.1.3 Lipopolysaccharide-induced*

LPS is another agent that has been extensively used to induce bladder-centric symptoms of cystitis either via intravesical instillation (0.1 mg/ml to 2 mg/ml) into the bladder of mice or rats for 30 to 45 minutes [81, 82, 84–88] or via intraperitoneal injection (20–25 mg/kg) [83, 84]. LPS is frequently installed into the bladder following PS to reduce the GAG layer, allowing LPS to directly bind to the urothelial cells [85–88]. LPS has been used to induce both acute cystitis by a single instillation [81, 83–86] and chronic cystitis by repeated instillations usually once a day for a total of four days [82] or once or twice a week for a total of five weeks [87, 88]. Instillation of LPS in rodents induces urothelial ulceration and hemorrhage [85], as well as bladder inflammation that persists up to five days after a single instillation. Severe disruption of bladder submucosal structures, urothelial denudation, edema, infiltration of mononuclear and polymorphonuclear leukocytes, as well as increased activity of myeloperoxidase with increased expression of proinflammatory cytokines IL1α, IL1β, IL6, IFN, TNFα, and NFκB were observed as early as 24 h after a single LPS instillation [81, 83–88]. At the same time point, increased urinary frequency, decreased intercontraction interval, mechanical allodynia, and pain-related behavior have also been observed [82, 84, 86, 88]. Chronic treatment with LPS causes severe damage to the urothelium, ultimately leading to tissue fibrosis and remodeling, characterized by increased expression of TGFβ [87]. Although LPS induces a phenotype similar to that observed in patients with IC/BPS, bacterial infection is an exclusion criterion for IC/BPS diagnosis, raising questions about the validity of this model.

### 5.2 Models with complex autoimmune mechanisms

Bladder-centric models successfully induce infiltration of innate immune cells, such as neutrophils, macrophages, and mast cells that are observed in patients with IC/BPS. However, recent studies have emphasized that an immune system dysfunction accompanied by clonal expansion of B-lymphocytes in the bladder might underlie the pathophysiology of HIC/BPS [25]. In accordance with the hypothesis of the autoimmune origin of IC/BPS, experimental autoimmune cystitis (EAC) models have been developed. Based on the method of inducement of bladder autoimmunity, EAC models can be subdivided into: (i) EAC models induced by bladder tissue homogenate; (ii) EAC models induced by uroplakins; and (iii) transgenic EAC models.

### *5.2.1 Bladder tissue homogenate-induced*

In these models, subcutaneous injection of homogenized bladder tissue is used to immunize recipient rodents, targeting the bladder with autoimmune mechanisms. Vaccinated animals develop bladder damage and inflammation characterized by infiltration and accumulation of CD4+ T cells and mast cells, along with increased expression of proinflammatory cytokines IL1β, IL4, IL6, IL10, IFNγ, and TNFα and chemokines CXCL9, CXCL10, and CXCL11. There are also increased urothelial permeability and signs of angiogenesis, indicated by increased expression of VCAM1, as well as glomerulations. These mice models exhibit functional changes such as increased urinary frequency, decreased bladder capacity, shorter intercontraction intervals, and reduced voided volumes per micturition. Additionally, mice develop hyperalgesia to von-Frey hair probing of the pelvic area and a decreased pelvic pain threshold [89–91, 110]. Furthermore, adoptive transfer of splenocytes cultured from the experimental animals to naïve recipients induced signs of cystitis in their bladders, including increased urinary frequency and vascular congestion [111].

### *5.2.2 Uroplakin-induced*

Uroplakins (UP) are a family of transmembrane proteins expressed on the terminally differentiated superficial urothelial cells. Four types of UP are currently known in humans—UPIa, UPIb, UPII, and UPIII. By forming urothelial plaques, they maintain barrier function and prevent passage of urine substances across the urothelium in the bladder [112]. In UP-induced models, a subcutaneous injection of recombinant mouse UP protein is used to trigger bladder inflammation. Several studies have reported that using UPII or UPIII antigens provokes infiltration of predominantly CD3+ T-lymphocytes and significantly elevates the expression of inflammatory cytokines IL1β, IL17, TNFα, and IFNγ within the bladder, along with the presence of autoantibodies in serum samples [93–97]. Bicer et al. noted higher expression of the chemokine CCL2 in the bladders of EAC mice, where CCL2 functions as a chemoattractant and activator of mast cells, which were also found in elevated numbers in the bladders of EAC mice [92]. UP EAC mice also exhibit altered bladder function characterized by voiding dysfunction, increased urinary frequency, decreased voided volume, and increased suprapubic pain [92–97]. Additionally, EAC mice injected with UP develop bladder tissue fibrosis, loss of urothelial barrier function (indicated by decreased expression of ZO1 and UPIII), and increased expression of toll-like receptors (TLR2, TLR4, TLR5, and TLR11) [95, 96]. In 2017, Song et al. compared five different experimental models of IC/BPS in rats, including transurethral instillation of HCl or acetic acid, intraperitoneal injection of CYP or LPS, and subcutaneous injection of UPII. Among these models, the injection of UP generated the most effective IC animal model, showing consequent urothelial barrier loss, inflammatory reaction, tissue fibrosis stimulation, and hyperactive bladder that persisted even 14 days after [113].

### *5.2.3 Transgenic URO/OVA models*

The most studied transgenic EAC model of IC/BPS is the URO/OVA mouse model. These mice express the membrane form of the model antigen ovalbumin (OVA) as a self-antigen on the urothelium. URO-OVA mice develop bladder inflammation at day 7 after adoptive transfer of activated OVA-specific CD8+ T cells from OT-I mice [36, 98]. Bladders from URO/OVA mice demonstrate severe inflammation with immune cell infiltration; edema; upregulated expression of IL6, TNFα, IFNγ, CCL2, NGF, and substance P; and urothelial hyperplasia 7–14 days after the transfer [98, 101, 114]. In addition, the URO-OVA mice exhibit increased pelvic nociceptive responses to von Frey filament stimulation and changes in voiding habits, including decreased mean voided volume and increased total number of voids [98, 99, 101, 114]. Bladder nociception in URO/OVA models is likely TLR4-dependent since URO/OVA-TLR4−/− mice exhibited significantly reduced pelvic and bladder nociceptive responses compared with wild-type URO-OVA mice after IC induction, although both developed similar bladder inflammation and voiding dysfunction [99]. Another transgenic EAC model was developed through crossbreeding between URO-OVA and OT-I mice. Bladder inflammation spontaneously develops at 4 weeks and persists up to 20 weeks. Similar to other EAC models, the URO/OVA/OT-I mice develop pelvic/bladder pain and voiding dysfunction, providing a valuable model for studying mechanisms of progression and chronicity of cystitis [98, 100, 101].

### 5.3 Psychological stress induced

Chronic psychological stress has recently been identified as a contributing factor in the development of IC/BPS. Studies have shown that individuals who experienced a traumatic event earlier in life are more likely to develop IC/BPS later in life [115, 116]. Mechanistically, stress results in chronic adrenergic stimulation, which induces pain and bladder changes. This process is mediated by alpha-1A adrenoceptors that enhance nociceptor responses, such as those from the transient receptor potential cation channel subfamily V member 1 (TRPV1), and increase ATP release upon bladder distension [117]. Patients with IC/BPS also exhibit increased sympathetic nervous system activity and elevated urinary noradrenaline levels [39]. To study the development of IC/BPS and replicate chronic stress, researchers have developed rodent models with induced psychological stress. The most commonly used stress inducer in these animal models is water avoidance stress (WAS).

#### *5.3.1 Water avoidance stress-induced*

The WAS model involves placing rodents on a small platform situated in the center of a tank filled with water. As rodents have a natural aversion to water and no escape route, this situation induces psychological stress and anxiety. Typically, chronic psychological stress in rodents is established after they are exposed to the WAS model for approximately one to two hours per day over a period of 5–10 days [102–106].

Rats exposed to WAS exhibit significant increase in urinary frequency and a reduced mechanical pain threshold. Additionally, bladders of WAS-exposed rats show infiltration of inflammatory cells and mastocytosis, as well as loss of superficial umbrella cells, resulting in an altered urothelial surface [105]. In a study by Zeybek et al., increased lipid peroxidation and decreased glutathione levels were observed in WAS-exposed rats, suggesting bladder injury caused by oxidative stress [102]. Furthermore, these rats exhibited decreased visceral hypersensitivity and stress-related cerebral activations within the supraspinal micturition circuit [103]. By day 8 of the WAS protocol, rats demonstrated a decreased pain threshold to mechanical stimulation, which persisted for up to one month [104]. Mice exposed to WAS exhibited increased urinary noradrenaline levels, infiltration of mast and lymphocytic cells, urothelial disruption, and increased urinary frequency evidenced by an increased mean number of voids and decreased voided volume per micturition [106].

WAS models have also been used as experimental models of irritable bowel syndrome [118], a chronic disorder which is a frequent comorbidity in patients with IC/BPS [15].

## 6. Conclusion

Preclinical studies using animal models remain the cornerstone of research into the etiology and pathophysiology of IC/BPS, the discovery of novel drug targets, and the design of future clinical trials. One of the key steps of animal models is to mimic as closely as possible all the major symptoms of IC/BPS. However, with various models employing different methods of cystitis induction, there is ongoing debate over which model is most appropriate for preclinical evaluation. Critical questions include whether a single model can encapsulate the disease's complexity or if multiple models are necessary, whether different models are required to study the various clinical subtypes of IC/BPS, and which model best addresses the chronicity of the disease. Additionally, the experimental design and quality of studies are crucial for successful translation into clinical settings. Our previous research indicated that the methodological quality of animal studies could be significantly enhanced, highlighting the need for measures to avoid bias to be implemented and properly reported [119]. Despite these concerns, substantial progress has been made in developing animal models of IC/BPS. These advancements have deepened our understanding of the complex mechanisms underlying IC/BPS pathophysiology and have led to the emergence of several novel treatment options.

## Acknowledgements

This work was financially supported by Slovenian Research and Innovation Agency (ARIS): project grant Z3-50118 to T.K. and National Research Programme P3-0108.

## Conflict of interest

The author declares no conflict of interest.

## Author details

Tadeja Kuret
Faculty of Medicine, Institute of Cell Biology, University of Ljubljana, Ljubljana, Slovenia

*Address all correspondence to: tadeja.kuret@mf.uni-lj.si

## References

[1] Homma Y, Ueda T, Tomoe H, Lin AT, Kuo HC, Lee MH, et al. Clinical guidelines for interstitial cystitis and hypersensitive bladder updated in 2015. International Journal of Urology. 2016;**23**(7):542-549

[2] Leppilahti M, Sairanen J, Tammela TL, Aaltomaa S, Lehtoranta K, Auvinen A, et al. Prevalence of clinically confirmed interstitial cystitis in women: A population based study in Finland. The Journal of Urology. 2005;**174**(2):581-583

[3] Choe JH, Son H, Song YS, Kim JC, Lee JZ, Lee KS. Prevalence of painful bladder syndrome/interstitial cystitis-like symptoms in women: A population-based study in Korea. World Journal of Urology. 2011;**29**(1):103-108

[4] Berry SH, Elliott MN, Suttorp M, Bogart LM, Stoto MA, Eggers P, et al. Prevalence of symptoms of bladder pain syndrome/interstitial cystitis among adult females in the United States. The Journal of Urology. 2011;**186**(2):540-544

[5] Clemens JQ, Erickson DR, Varela NP, Lai HH. Diagnosis and treatment of interstitial cystitis/bladder pain syndrome. The Journal of Urology. 2022;**208**(1):34-42

[6] Jhang JF, Jiang YH, Kuo HC. Current understanding of the pathophysiology and novel treatments of interstitial cystitis/bladder pain syndrome. Biomedicines. 2022;**10**(10):2380. DOI: 10.3390/biomedicines10102380

[7] Colemeadow J, Sahai A, Malde S. Clinical Management of Bladder Pain Syndrome/interstitial cystitis: A review on current recommendations and emerging treatment options. Research and Reports in Urology. 2020;**12**:331-343

[8] Akiyama Y, Luo Y, Hanno PM, Maeda D, Homma Y. Interstitial cystitis/bladder pain syndrome: The evolving landscape, animal models and future perspectives. International Journal of Urology. 2020;**27**(6):491-503

[9] Birder L, Andersson KE. Animal modelling of interstitial cystitis/bladder pain syndrome. International Neurourology Journal. 2018;**22**(Suppl. 1):S3-S9

[10] Juliebo-Jones P, Hjelle KM, Mohn J, Gudbrandsdottir G, Roth I, Chaudhry AA, et al. Management of Bladder Pain Syndrome (BPS): A practical guide. Advances in Urology. 2022;**2022**:7149467

[11] Akiyama Y, Hanno P. Phenotyping of interstitial cystitis/bladder pain syndrome. International Journal of Urology. 2019;**26**(Suppl. 1):17-19

[12] Homma Y, Akiyama Y, Tomoe H, Furuta A, Ueda T, Maeda D, et al. Clinical guidelines for interstitial cystitis/bladder pain syndrome. International Journal of Urology. 2020;**27**(7):578-589

[13] Jhang JF, Kuo HC. Pathomechanism of interstitial cystitis/bladder pain syndrome and mapping the heterogeneity of disease. International Neurourology Journal. 2016;**20**(Suppl. 2):S95-S104

[14] Yueh HZ, Yang MH, Huang JY, Wei JC. Risk of autoimmune diseases in patients with interstitial cystitis/bladder pain syndrome: A Nationwide population-based study in Taiwan. Frontiers in Medicine (Lausanne). 2021;**8**:747098

[15] Chelimsky G, Heller E, Buffington CA, Rackley R, Zhang D,

Chelimsky T. Co-morbidities of interstitial cystitis. Frontiers in Neuroscience. 2012;**6**:114

[16] Wei B, Zhao Y, Lin P, Qiu W, Wang S, Gu C, et al. The association between overactive bladder and systemic immunity-inflammation index: A cross-sectional study of NHANES 2005 to 2018. Scientific Reports. 2024;**14**(1):12579

[17] Grundy L, Caldwell A, Brierley SM. Mechanisms underlying overactive bladder and interstitial cystitis/painful bladder syndrome. Frontiers in Neuroscience. 2018;**12**:931

[18] Jafari NV, Rohn JL. The urothelium: A multi-faceted barrier against a harsh environment. Mucosal Immunology. 2022;**15**(6):1127-1142

[19] Hurst RE, Greenwood-Van Meerveld B, Wisniewski AB, VanGordon S, Lin H, Kropp BP, et al. Increased bladder permeability in interstitial cystitis/painful bladder syndrome. Translational Andrology and Urology. 2015;**4**(5):563-571

[20] Liu HT, Shie JH, Chen SH, Wang YS, Kuo HC. Differences in mast cell infiltration, E-cadherin, and zonula occludens-1 expression between patients with overactive bladder and interstitial cystitis/bladder pain syndrome. Urology. 2012;**80**(1):225 e13-225 e18

[21] Lee JD, Lee MH. Decreased expression of zonula occludens-1 and occludin in the bladder urothelium of patients with interstitial cystitis/painful bladder syndrome. Journal of the Formosan Medical Association. 2014;**113**(1):17-22

[22] Cho KJ, Lee KS, Choi JB, Koh JS, Kim JC. Changes in uroplakin expression in the urothelium of patients with ulcerative interstitial cystitis/bladder pain syndrome. Investigative and Clinical Urology. 2020;**61**(3):304-309

[23] Lee YK, Jhang JF, Jiang YH, Hsu YH, Ho HC, Kuo HC. Difference in electron microscopic findings among interstitial cystitis/bladder pain syndrome with distinct clinical and cystoscopic characteristics. Scientific Reports. 2021;**11**(1):17258

[24] Keay S. Cell signaling in interstitial cystitis/painful bladder syndrome. Cellular Signalling. 2008;**20**(12):2174-2179

[25] Moldwin RM, Nursey V, Yaskiv O, Dalvi S, Macdonald EJ, Funaro M, et al. Immune cell profiles of patients with interstitial cystitis/bladder pain syndrome. Journal of Translational Medicine. 2022;**20**(1):97

[26] Peng L, Jin X, Li BY, Zeng X, Liao BH, Jin T, et al. Integrating single-cell RNA sequencing with spatial transcriptomics reveals immune landscape for interstitial cystitis. Signal Transduction and Targeted Therapy. 2022;**7**(1):161

[27] Wang ZY, Bjorling DE. Tumor necrosis factor-alpha induces expression and release of interleukin-6 by human urothelial cells. Inflammation Research. 2011;**60**(6):525-532

[28] Kuret T, Peskar D, Kreft ME, Erman A, Veranic P. Comprehensive transcriptome profiling of urothelial cells following TNFalpha stimulation in an in vitro interstitial cystitis/bladder pain syndrome model. Frontiers in Immunology. 2022;**13**:960667

[29] Jin XW, Wang QZ, Zhao Y, Liu BK, Zhang X, Wang XJ, et al. An experimental model of the epithelial to mesenchymal transition and pro-fibrogenesis in urothelial cells related

to bladder pain syndrome/interstitial cystitis. Translational Andrology and Urology. 2021;**10**(11):4120-4131

[30] Akiyama Y. Biomarkers in interstitial cystitis/bladder pain syndrome with and without Hunner lesion: A review and future perspectives. Diagnostics (Basel). 2021;**11**(12):2238. DOI: 10.3390/diagnostics11122238

[31] Jiang YH, Jhang JF, Kuo HC. Can we use urinary cytokine/chemokine analysis in discriminating ulcer-type interstitial cystitis/bladder pain syndrome? Diagnostics (Basel). 2022;**12**(5):1093. DOI: 10.3390/diagnostics12051093

[32] Jiang YH, Peng CH, Liu HT, Kuo HC. Increased pro-inflammatory cytokines, C-reactive protein and nerve growth factor expressions in serum of patients with interstitial cystitis/bladder pain syndrome. PLoS One. 2013;**8**(10):e76779

[33] Chung SD, Liu HT, Lin H, Kuo HC. Elevation of serum c-reactive protein in patients with OAB and IC/BPS implies chronic inflammation in the urinary bladder. Neurourology and Urodynamics. 2011;**30**(3):417-420

[34] Ochs RL, Stein TW Jr, Peebles CL, Gittes RF, Tan EM. Autoantibodies in interstitial cystitis. The Journal of Urology. 1994;**151**(3):587-592

[35] van de Merwe JP. Interstitial cystitis and systemic autoimmune diseases. Nature Clinical Practice. Urology. 2007;**4**(9):484-491

[36] Akiyama Y, Yao JR, Kreder KJ, O'Donnell MA, Lutgendorf SK, Lyu D, et al. Autoimmunity to urothelial antigen causes bladder inflammation, pelvic pain, and voiding dysfunction: A novel animal model for Hunner-type interstitial cystitis. American Journal of Physiology. Renal Physiology. 2021;**320**(2):F174-FF82

[37] Nickel JC, Tripp DA, Pontari M, Moldwin R, Mayer R, Carr LK, et al. Childhood sexual trauma in women with interstitial cystitis/bladder pain syndrome: A case control study. Canadian Urological Association Journal. 2011;**5**(6):410-415

[38] Laden BF, Bresee C, De Hoedt A, Dallas KB, Scharfenberg A, Saxena R, et al. Comorbidities in a Nationwide, Heterogenous population of veterans with interstitial cystitis/bladder pain syndrome. Urology. 2021;**156**:37-43

[39] Lutgendorf SK, Zia S, Luo Y, O'Donnell M, van Bokhoven A, Bradley CS, et al. Early and recent exposure to adversity, TLR-4 stimulated inflammation, and diurnal cortisol in women with interstitial cystitis/bladder pain syndrome: A MAPP research network study. Brain, Behavior, and Immunity. 2023;**111**:116-123

[40] Tay C, Grundy L. Animal models of interstitial cystitis/bladder pain syndrome. Frontiers in Physiology. 2023;**14**:1232017

[41] Boudieu L, Mountadem S, Lashermes A, Meleine M, Ulmann L, Rassendren F, et al. Blocking alpha(2) delta-1 subunit reduces bladder hypersensitivity and inflammation in a cystitis mouse model by decreasing NF-kB pathway activation. Frontiers in Pharmacology. 2019;**10**:133

[42] Zhang X, Gao S, Tanaka M, Zhang Z, Huang Y, Mitsui T, et al. Carbenoxolone inhibits TRPV4 channel-initiated oxidative urothelial injury and ameliorates cyclophosphamide-induced bladder dysfunction. Journal of Cellular and Molecular Medicine. 2017;**21**(9):1791-1802

[43] Liu Q, Wu Z, Liu Y, Chen L, Zhao H, Guo H, et al. Cannabinoid receptor 2 activation decreases severity of cyclophosphamide-induced cystitis via regulating autophagy. Neurourology and Urodynamics. 2020;**39**(1):158-169

[44] de Oliveira MG, Calmasini FB, Alexandre EC, De Nucci G, Monica FZ, Antunes E. Activation of soluble guanylyl cyclase by BAY 58-2667 improves bladder function in cyclophosphamide-induced cystitis in mice. American Journal of Physiology. Renal Physiology. 2016;**311**(1):F85-F93

[45] Chen YT, Yang CC, Sun CK, Chiang HJ, Chen YL, Sung PH, et al. Extracorporeal shock wave therapy ameliorates cyclophosphamide-induced rat acute interstitial cystitis though inhibiting inflammation and oxidative stress-in vitro and in vivo experiment studies. American Journal of Translational Research. 2014;**6**(6):631-648

[46] Smaldone MC, Vodovotz Y, Tyagi V, Barclay D, Philips BJ, Yoshimura N, et al. Multiplex analysis of urinary cytokine levels in rat model of cyclophosphamide-induced cystitis. Urology. 2009;**73**(2):421-426

[47] Chopra B, Barrick SR, Meyers S, Beckel JM, Zeidel ML, Ford AP, et al. Expression and function of bradykinin B1 and B2 receptors in normal and inflamed rat urinary bladder urothelium. The Journal of Physiology. 2005;**562**(Pt 3):859-871

[48] Liu M, Shen S, Kendig DM, Mahavadi S, Murthy KS, Grider JR, et al. Inhibition of NMDAR reduces bladder hypertrophy and improves bladder function in cyclophosphamide induced cystitis. The Journal of Urology. 2015;**193**(5):1676-1683

[49] Chen YH, Man KM, Chen WC, Liu PL, Tsai KS, Tsai MY, et al. Platelet-rich plasma ameliorates cyclophosphamide-induced acute interstitial cystitis/painful bladder syndrome in a rat model. Diagnostics (Basel). 2020;**10**(6):381. DOI: 10.3390/diagnostics10060381

[50] Coelho A, Wolf-Johnston AS, Shinde S, Cruz CD, Cruz F, Avelino A, et al. Urinary bladder inflammation induces changes in urothelial nerve growth factor and TRPV1 channels. British Journal of Pharmacology. 2015;**172**(7):1691-1699

[51] Auge C, Chene G, Dubourdeau M, Desoubzdanne D, Corman B, Palea S, et al. Relevance of the cyclophosphamide-induced cystitis model for pharmacological studies targeting inflammation and pain of the bladder. European Journal of Pharmacology. 2013;**707**(1-3):32-40

[52] Pessina F, Capasso R, Borrelli F, Aveta T, Buono L, Valacchi G, et al. Protective effect of palmitoy-lethanolamide in a rat model of cystitis. The Journal of Urology. 2015;**193**(4):1401-1408

[53] Juszczak K, Gil K, Wyczolkowski M, Thor PJ. Functional, histological structure and mastocytes alterations in rat urinary bladders following acute and [corrected] chronic cyclophosphamide treatment. Journal of Physiology and Pharmacology. 2010;**61**(4):477-482

[54] Malley SE, Vizzard MA. Changes in urinary bladder cytokine mRNA and protein after cyclophosphamide-induced cystitis. Physiological Genomics. 2002;**9**(1):5-13

[55] DeBerry JJ, Saloman JL, Dragoo BK, Albers KM, Davis BM. Artemin immunotherapy is effective

in preventing and reversing cystitis-induced bladder hyperalgesia via TRPA1 regulation. The Journal of Pain. 2015;**16**(7):628-636

[56] Boudes M, Uvin P, Kerselaers S, Vennekens R, Voets T, De Ridder D. Functional characterization of a chronic cyclophosphamide-induced overactive bladder model in mice. Neurourology and Urodynamics. 2011;**30**(8):1659-1665

[57] Golubeva AV, Zhdanov AV, Mallel G, Dinan TG, Cryan JF. The mouse cyclophosphamide model of bladder pain syndrome: Tissue characterization, immune profiling, and relationship to metabotropic glutamate receptors. Physiological Reports. 2014;**2**(3):e00260

[58] Peskar D, Kuret T, Lakota K, Erman A. Molecular profiling of inflammatory processes in a mouse model of IC/BPS: From the complete transcriptome to major sex-related histological features of the urinary bladder. International Journal of Molecular Sciences. 2023;**24**(6):5758. DOI: 10.3390/ijms24065758

[59] Auge C, Game X, Vergnolle N, Lluel P, Chabot S. Characterization and validation of a chronic model of cyclophosphamide-induced interstitial cystitis/bladder pain syndrome in rats. Frontiers in Pharmacology. 2020;**11**:1305

[60] Mahal A, Young-Lin N, Dobberfuhl A, Estes J, Comiter CV. Peroxisome proliferator-activated receptor gamma agonist as a novel treatment for interstitial cystitis: A rat model. Investigative and Clinical Urology. 2018;**59**(4):257-262

[61] Wang HJ, Lee WC, Tyagi P, Huang CC, Chuang YC. Effects of low energy shock wave therapy on inflammatory moleculars, bladder pain, and bladder function in a rat cystitis model. Neurourology and Urodynamics. 2017;**36**(6):1440-1447

[62] Li W, Yang F, Zhan H, Liu B, Cai J, Luo Y, et al. Houttuynia cordata extract ameliorates bladder damage and improves bladder symptoms via anti-inflammatory effect in rats with interstitial cystitis. Evidence-based Complementary and Alternative Medicine. 2020;**2020**:9026901

[63] Luo J, Yang C, Luo X, Yang Y, Li J, Song B, et al. Chlorogenic acid attenuates cyclophosphamide-induced rat interstitial cystitis. Life Sciences. 2020;**254**:117590

[64] Xie J, Liu B, Chen J, Xu Y, Zhan H, Yang F, et al. Umbilical cord-derived mesenchymal stem cells alleviated inflammation and inhibited apoptosis in interstitial cystitis via AKT/mTOR signaling pathway. Biochemical and Biophysical Research Communications. 2018;**495**(1):546-552

[65] Yang Y, Zhang H, Lu Q, Liu X, Fan Y, Zhu J, et al. Suppression of adenosine a(2a) receptors alleviates bladder overactivity and hyperalgesia in cyclophosphamide-induced cystitis by inhibiting TRPV1. Biochemical Pharmacology. 2021;**183**:114340

[66] Ko IG, Jin JJ, Hwang L, Kim SH, Kim CJ, Won KY, et al. Adenosine a(2A) receptor agonist Polydeoxyribonucleotide alleviates interstitial cystitis-induced voiding dysfunction by suppressing inflammation and apoptosis in rats. Journal of Inflammation Research. 2021;**14**:367-378

[67] Funahashi Y, Yoshida M, Yamamoto T, Majima T, Takai S, Gotoh M. Intravesical application of rebamipide promotes urothelial healing in a rat cystitis model. The Journal of Urology. 2014;**192**(6):1864-1870

[68] Song M, Lim J, Yu HY, Park J, Chun JY, Jeong J, et al. Mesenchymal stem cell therapy alleviates interstitial cystitis by activating Wnt Signaling pathway. Stem Cells and Development. 2015;**24**(14):1648-1657

[69] Kim A, Yu HY, Lim J, Ryu CM, Kim YH, Heo J, et al. Improved efficacy and in vivo cellular properties of human embryonic stem cell derivative in a preclinical model of bladder pain syndrome. Scientific Reports. 2017;**7**(1):8872

[70] Shimizu N, De Velasco MA, Umekawa T, Uemura H, Yoshikawa K. Effects of the rho kinase inhibitor, hydroxyfasudil, on bladder dysfunction and inflammation in rats with HCl-induced cystitis. International Journal of Urology. 2013;**20**(11):1136-1143

[71] Konkol Y, Bernoulli J, Streng T, Jaaskelainen K, Laihia J, Leino L. Intravesical treatment with cis-urocanic acid improves bladder function in rat model of acute bladder inflammation. Neurourology and Urodynamics. 2016;**35**(7):786-791

[72] Hirose Y, Yamamoto T, Nakashima M, Funahashi Y, Matsukawa Y, Yamaguchi M, et al. Injection of dental pulp stem cells promotes healing of damaged bladder tissue in a rat model of chemically induced cystitis. Cell Transplantation. 2016;**25**(3):425-436

[73] Furuta A, Yamamoto T, Igarashi T, Suzuki Y, Egawa S, Yoshimura N. Bladder wall injection of mesenchymal stem cells ameliorates bladder inflammation, overactivity, and nociception in a chemically induced interstitial cystitis-like rat model. International Urogynecology Journal. 2018;**29**(11):1615-1622

[74] Tyagi P, Chancellor M, Yoshimura N, Huang L. Activity of different phospholipids in attenuating hyperactivity in bladder irritation. BJU International. 2008;**101**(5):627-632

[75] Tyagi P, Hsieh VC, Yoshimura N, Kaufman J, Chancellor MB. Instillation of liposomes vs dimethyl sulphoxide or pentosan polysulphate for reducing bladder hyperactivity. BJU International. 2009;**104**(11):1689-1692

[76] Fraser MO, Chuang YC, Tyagi P, Yokoyama T, Yoshimura N, Huang L, et al. Intravesical liposome administration--a novel treatment for hyperactive bladder in the rat. Urology. 2003;**61**(3):656-663

[77] Soler R, Bruschini H, Truzzi JC, Martins JR, Camara NO, Alves MT, et al. Urinary glycosaminoglycans excretion and the effect of dimethyl sulfoxide in an experimental model of non-bacterial cystitis. International Brazilian Journal of Urology. 2008;**34**(4):503-511; discussion 11

[78] Cetinel S, Canillioglu YE, Cikler E, Sener G, Ercan F. Leukotriene D4 receptor antagonist montelukast alleviates protamine sulphate-induced changes in rat urinary bladder. BJU International. 2011;**107**(8):1320-1325

[79] Akin Y, Bozkurt A, Erol HS, Halici M, Celebi F, Kapakin KA, et al. Impact of rho-kinase inhibitor Hydroxyfasudil in protamine sulphate induced cystitis rat bladder. Low Urinary Tract Symptoms. 2015;7(2):108-114

[80] Grundy L, Caldwell A, Lumsden A, Mohammadi E, Hannig G, Greenwood Van-Meervald B, et al. Experimentally induced bladder permeability evokes bladder afferent hypersensitivity in the absence of inflammation. Frontiers in Neuroscience. 2020;**14**:590871

[81] Gonzalez RR, Fong T, Belmar N, Saban M, Felsen D, Te A. Modulating

bladder neuro-inflammation: RDP58, a novel anti-inflammatory peptide, decreases inflammation and nerve growth factor production in experimental cystitis. The Journal of Urology. 2005;**173**(2):630-634

[82] Yoshizumi M, Watanabe C, Mizoguchi H. Gabapentin reduces painful bladder hypersensitivity in rats with lipopolysaccharide-induced chronic cystitis. Pharmacology Research & Perspectives. 2021;**9**(1):e00697

[83] Tambaro S, Casu MA, Mastinu A, Lazzari P. Evaluation of selective cannabinoid CB(1) and CB(2) receptor agonists in a mouse model of lipopolysaccharide-induced interstitial cystitis. European Journal of Pharmacology. 2014;**729**:67-74

[84] Berger G, Arora N, Burkovskiy I, Xia Y, Chinnadurai A, Westhofen R, et al. Experimental cannabinoid 2 receptor activation by Phyto-derived and synthetic cannabinoid ligands in LPS-induced interstitial cystitis in mice. Molecules. 2019;**24**(23):4239. DOI: 10.3390/molecules24234239

[85] Li J, Luo H, Dong X, Liu Q, Wu C, Zhang T, et al. Therapeutic effect of urine-derived stem cells for protamine/lipopolysaccharide-induced interstitial cystitis in a rat model. Stem Cell Research & Therapy. 2017;**8**(1):107

[86] Song YJ, Cao JY, Jin Z, Hu WG, Wu RH, Tian LH, et al. Inhibition of microRNA-132 attenuates inflammatory response and detrusor fibrosis in rats with interstitial cystitis via the JAK-STAT signaling pathway. Journal of Cellular Biochemistry. 2019;**120**(6):9147-9158

[87] Shih HJ, Chang CY, Lai CH, Huang CJ. Therapeutic effect of modulating the NLRP3-regulated transforming growth factor-beta signaling pathway on interstitial cystitis/bladder pain syndrome. Biomedicine & Pharmacotherapy. 2021;**138**:111522

[88] Ryu CM, Yu HY, Lee HY, Shin JH, Lee S, Ju H, et al. Longitudinal intravital imaging of transplanted mesenchymal stem cells elucidates their functional integration and therapeutic potency in an animal model of interstitial cystitis/bladder pain syndrome. Theranostics. 2018;**8**(20):5610-5624

[89] Liu BK, Jin XW, Lu HZ, Zhang X, Zhao ZH, Shao Y. The effects of Neurokinin-1 receptor antagonist in an experimental autoimmune cystitis model resembling bladder pain syndrome/interstitial cystitis. Inflammation. 2019;**42**(1):246-254

[90] Singh UP, Singh NP, Guan H, Hegde VL, Price RL, Taub DD, et al. The severity of experimental autoimmune cystitis can be ameliorated by anti-CXCL10 ab treatment. PLoS One. 2013;**8**(11):e79751

[91] Lin YH, Liu G, Kavran M, Altuntas CZ, Gasbarro G, Tuohy VK, et al. Lower urinary tract phenotype of experimental autoimmune cystitis in mouse: A potential animal model for interstitial cystitis. BJU International. 2008;**102**(11):1724-1730

[92] Bicer F, Altuntas CZ, Izgi K, Ozer A, Kavran M, Tuohy VK, et al. Chronic pelvic allodynia is mediated by CCL2 through mast cells in an experimental autoimmune cystitis model. American Journal of Physiology. Renal Physiology. 2015;**308**(2):F103-F113

[93] Li H, Zhang Z, Peng J, Xin Z, Li M, Yang B, et al. Treatment with low-energy shock wave alleviates pain in an animal model of uroplakin 3A-induced autoimmune interstitial cystitis/painful

bladder syndrome. Investigative and Clinical Urology. 2019;**60**(5):359-366

[94] Izgi K, Altuntas CZ, Bicer F, Ozer A, Sakalar C, Li X, et al. Uroplakin peptide-specific autoimmunity initiates interstitial cystitis/painful bladder syndrome in mice. PLoS One. 2013;**8**(8):e72067

[95] Kim BS, Chun SY, Lee EH, Chung JW, Lee JN, Ha YS, et al. Efficacy of combination therapy with pentosan polysulfate sodium and adipose tissue-derived stem cells for the management of interstitial cystitis in a rat model. Stem Cell Research. 2020;**45**:101801

[96] Chung JW, Chun SY, Lee EH, Ha YS, Lee JN, Song PH, et al. Verification of mesenchymal stem cell injection therapy for interstitial cystitis in a rat model. PLoS One. 2019;**14**(12):e0226390

[97] Altuntas CZ, Daneshgari F, Sakalar C, Goksoy E, Gulen MF, Kavran M, et al. Autoimmunity to uroplakin II causes cystitis in mice: A novel model of interstitial cystitis. European Urology. 2012;**61**(1):193-200

[98] Liu W, Evanoff DP, Chen X, Luo Y. Urinary bladder epithelium antigen induces CD8+ T cell tolerance, activation, and autoimmune response. Journal of Immunology. 2007;**178**(1):539-546

[99] Cui X, Jing X, Lutgendorf SK, Bradley CS, Schrepf A, Erickson BA, et al. Cystitis-induced bladder pain is toll-like receptor 4 dependent in a transgenic autoimmune cystitis murine model: A MAPP research network animal study. American Journal of Physiology. Renal Physiology. 2019;**317**(1):F90-FF8

[100] Kogan P, Xu S, Wang Y, O'Donnell MA, Lutgendorf SK, Bradley CS, et al. Sub-noxious Intravesical lipopolysaccharide triggers bladder inflammation and symptom onset in a transgenic autoimmune cystitis model: A MAPP network animal study. Scientific Reports. 2018;**8**(1):6573

[101] Kim R, Liu W, Chen X, Kreder KJ, Luo Y. Intravesical dimethyl sulfoxide inhibits acute and chronic bladder inflammation in transgenic experimental autoimmune cystitis models. Journal of Biomedicine & Biotechnology. 2011;**2011**:937061

[102] Zeybek A, Saglam B, Cikler E, Cetinel S, Ercan F, Sener G. Taurine ameliorates stress-induced degeneration of the urinary bladder. Acta Histochemica. 2007;**109**(3):208-214

[103] Holschneider DP, Wang Z, Chang H, Zhang R, Gao Y, Guo Y, et al. Ceftriaxone inhibits stress-induced bladder hyperalgesia and alters cerebral micturition and nociceptive circuits in the rat: A multidisciplinary approach to the study of urologic chronic pelvic pain syndrome research network study. Neurourology and Urodynamics. 2020;**39**(6):1628-1643

[104] Lee UJ, Ackerman AL, Wu A, Zhang R, Leung J, Bradesi S, et al. Chronic psychological stress in high-anxiety rats induces sustained bladder hyperalgesia. Physiology & Behavior. 2015;**139**:541-548

[105] Bazi T, Hajj-Hussein IA, Awwad J, Shams A, Hijaz M, Jurjus A. A modulating effect of epigallocatechin gallate (EGCG), a tea catechin, on the bladder of rats exposed to water avoidance stress. Neurourology and Urodynamics. 2013;**32**(3):287-292

[106] West EG, Sellers DJ, Chess-Williams R, McDermott C. Bladder overactivity induced by psychological stress in female mice is associated with

enhanced bladder contractility. Life Sciences. 2021;**265**:118735

[107] Mills KA, Chess-Williams R, McDermott C. Novel insights into the mechanism of cyclophosphamide-induced bladder toxicity: chloroacetaldehyde's contribution to urothelial dysfunction in vitro. Archives of Toxicology. 2019;**93**(11):3291-3303

[108] Greenwood-Van Meerveld B, Mohammadi E, Latorre R, Truitt ER 3rd, Jay GD, Sullivan BD, et al. Preclinical animal studies of Intravesical recombinant human proteoglycan 4 as a novel potential therapy for diseases resulting from increased bladder permeability. Urology. 2018;**116**(230):e1-e7

[109] Lavelle J, Meyers S, Ramage R, Bastacky S, Doty D, Apodaca G, et al. Bladder permeability barrier: Recovery from selective injury of surface epithelial cells. American Journal of Physiology. Renal Physiology. 2002;**283**(2):F242-F253

[110] Jin XW, Liu BK, Zhang X, Zhao ZH, Shao Y. Establishment of a novel autoimmune experimental model of bladder pain syndrome/interstitial cystitis in C57BL/6 mice. Inflammation. 2017;**40**(3):861-870

[111] Luber-Narod J, Austin-Ritchie T, Banner B, Hollins C 3rd, Maramag C, Price H, et al. Experimental autoimmune cystitis in the Lewis rat: A potential animal model for interstitial cystitis. Urological Research. 1996;**24**(6):367-373

[112] Sivakumaar K, Griffin J, Schofield E, Catto JWF, Jubber I. Gene of the month: The uroplakins. Journal of Clinical Pathology. 2024;**77**(5):291-296

[113] Song PH, Chun SY, Chung JW, Kim YY, Lee HJ, Lee JN, et al. Comparison of 5 different rat models to establish a standard animal model for research into interstitial cystitis. International Neurourology Journal. 2017;**21**(3):163-170

[114] Liu W, Deyoung BR, Chen X, Evanoff DP, Luo Y. RDP58 inhibits T cell-mediated bladder inflammation in an autoimmune cystitis model. Journal of Autoimmunity. 2008;**30**(4):257-265

[115] Schrepf A, Naliboff B, Williams DA, Stephens-Shields AJ, Landis JR, Gupta A, et al. Adverse childhood experiences and symptoms of urologic chronic pelvic pain syndrome: A multidisciplinary approach to the study of chronic pelvic pain research network study. Annals of Behavioral Medicine. 2018;**52**(10):865-877

[116] Gupta A, Bhatt RR, Naliboff BD, Kutch JJ, Labus JS, Vora PP, et al. Impact of early adverse life events and sex on functional brain networks in patients with urological chronic pelvic pain syndrome (UCPPS): A MAPP research network study. PLoS One. 2019;**14**(6):e0217610

[117] Charrua A, Pinto R, Birder LA, Cruz F. Sympathetic nervous system and chronic bladder pain: A new tune for an old song. Translational Andrology and Urology. 2015;**4**(5):534-542

[118] Vannucchi MG, Evangelista S. Experimental models of irritable bowel syndrome and the role of the enteric neurotransmission. Journal of Clinical Medicine. 2018;7(1):1

[119] Kuret T, Peskar D, Erman A, Veranic P. A systematic review of therapeutic approaches used in experimental models of interstitial cystitis/bladder pain syndrome. Biomedicine. 2021;**9**(8):865. DOI: 10.3390/biomedicines9080865

## Chapter 6

# Features of Anesthesia of Small and Large Laboratory Animals

*Petr Peretyagin and Anna Soloveva*


## Abstract

In world practice, conducting biological, pharmacological, medical, and a number of other studies is often impossible without animal experiments. According to the provisions of the "European Convention for the Protection of Vertebrate Animals Used for Experimental and Other Scientific Purposes," as well as the code of good laboratory practice (GLP) for animal surgery, depending on the experimental conditions, an important aspect is the anesthetic manual. The speed of recovery of the animal and, as a result, the final result of the experiment depends on the quality of anesthesia and subsequent analgesia. Each animal is "worth its weight in gold." This is true for both rats and rabbits. Especially when working with large animals (pigs, mini piggies, and sheep), taking into account the cost and features of maintenance, the choice of drugs and anesthesia regimens during surgery, as well as support in the postoperative period, become the most relevant. In modern literature, this issue is covered; however, there are practical nuances that cannot be found in the literature since they are revealed only in practice. For a successful experiment, knowledge of such aspects will help to save the lives of animals from the first days of the experiment. It is precisely these features that this chapter will be devoted to.


**Keywords:** anesthesia, experiment, laboratory animals, sedation, experimental operations

## 1. Introduction

Anesthesia plays a critical role in the realm of laboratory animal research. It is an essential component that ensures the humane treatment of animals during experimental procedures, surgical interventions, and diagnostic assessments. The application of anesthesia not only alleviates pain and distress in animals but also contributes to the accuracy and reliability of scientific data by minimizing physiological stress responses. This book aims to provide a comprehensive understanding of the features, techniques, and considerations involved in administering anesthesia to both small and large laboratory animals [1–3].

The use of anesthesia in laboratory animals is governed by stringent ethical guidelines and regulatory standards designed to uphold animal welfare. Ethical considerations dictate that any research involving animals must prioritize their well-being, employing anesthesia to prevent suffering and distress. Researchers and veterinarians are entrusted with the responsibility of ensuring that anesthetic protocols are appropriately designed and meticulously executed to meet the highest standards of care.

This not only enhances the ethical integrity of the research but also reinforces public trust in scientific endeavors [4–6].

The evolution of anesthesia in laboratory animals mirrors the advancements in human medicine. Early experiments were often conducted without adequate pain relief, reflecting a limited understanding of animal physiology and welfare. However, with the progression of veterinary science and the growing emphasis on ethical research practices, significant strides have been made in the development of safe and effective anesthetic techniques for various animal species. Pioneering work in this field has led to the refinement of anesthetic agents and the introduction of innovative methods, ensuring the humane treatment of laboratory animals and the reliability of experimental outcomes [7–9].

Anesthesia is a complex physiological state induced by administering specific agents that produce reversible loss of sensation and consciousness. The mechanisms underlying anesthesia involve the interaction of anesthetic drugs with the central nervous system, leading to the inhibition of neuronal activity and the modulation of synaptic transmission. Understanding these mechanisms is fundamental to selecting appropriate anesthetic protocols and managing the depth of anesthesia required for various procedures [10, 11].

Anesthetic agents can be broadly categorized into inhalational and injectable types, each with distinct pharmacological properties and applications. Inhalational agents, such as isoflurane and sevoflurane, are administered *via* inhalation and are known for their rapid onset and easy adjustability. Injectable agents, including ketamine and propofol, are administered through intravenous or intramuscular routes, offering advantages in terms of precise dosing and ease of use. The choice of anesthetic agent depends on factors such as the species, the procedure, and the duration of anesthesia required [12, 13].

The administration of anesthesia is typically divided into stages and planes, which describe the depth of anesthesia and its progression. These stages range from the initial induction, characterized by a gradual loss of consciousness, to the maintenance phase, where the desired level of anesthesia is sustained. Understanding the stages and planes of anesthesia is crucial for monitoring the anesthetic depth and ensuring the safety and well-being of the animal throughout the procedure. Proper monitoring techniques and equipment are essential to detect any deviations from the desired anesthetic plane and to make necessary adjustments promptly [14, 15].

This introduction sets the stage for a detailed exploration of anesthesia in laboratory animals, laying the groundwork for the comprehensive and nuanced discussions that follow in subsequent chapters.

## 2. Anesthesia in small laboratory animals

The administration of anesthesia in small laboratory animals presents unique challenges and considerations due to the diverse physiological characteristics and varying responses to anesthetic agents among different species. This section delves into the specific techniques, protocols, and care required for effectively anesthetizing small laboratory animals, ensuring their welfare while facilitating accurate and reliable scientific research [16–18].

## 2.1 Rodents (mice and rats)

Rodents, particularly mice and rats, are among the most commonly used animals in biomedical research. Their small size, high metabolic rate, and specific physiological traits necessitate tailored anesthetic techniques to achieve effective anesthesia while minimizing stress and potential complications.

Anesthetic techniques: Various methods are employed to induce anesthesia in rodents, including inhalational anesthesia with agents like isoflurane and injectable anesthesia with agents such as ketamine-xylazine combinations. Each technique requires careful consideration of dosage, route of administration, and the individual animal's condition.

Commonly used agents: Isoflurane, ketamine, and xylazine are frequently used due to their efficacy and safety profiles. The choice of agent depends on the type of procedure, the duration of anesthesia required, and the specific needs of the study.

Monitoring and recovery: Continuous monitoring of vital signs, such as respiratory rate, heart rate, and body temperature, is essential to ensure the animal's well-being during anesthesia. Postoperative care focuses on providing a warm, quiet environment to facilitate recovery and prevent complications.

## 2.2 Rabbits

Rabbits require specialized anesthetic protocols due to their unique anatomical and physiological characteristics, such as their relatively high body surface area to volume ratio and sensitivity to stress.

Preanesthetic preparation: Proper fasting, hydration, and preanesthetic assessment are crucial steps to minimize risks and ensure a smooth induction of anesthesia. Sedation may be used to reduce stress and facilitate handling.

Anesthetic protocols: Injectable agents like ketamine combined with medetomidine or inhalational agents such as isoflurane are commonly used. Protocols are tailored to the specific needs of the procedure and the individual rabbit.

Postoperative care: Postoperative care for rabbits involves monitoring for hypothermia, providing analgesia to manage pain, and ensuring adequate hydration and nutrition during the recovery period.

## 2.3 Guinea pigs

Guinea pigs present particular anesthetic challenges due to their unique anatomical features and sensitivity to respiratory depressants. Careful selection of anesthetic agents and techniques is critical to ensuring effective anesthesia and minimizing complications.

Anesthetic considerations: Factors such as the animal's size, age, and health status must be taken into account when planning anesthesia. Adequate preanesthetic fasting and hydration are important to reduce the risk of aspiration and other complications.

Agent selection and administration: Isoflurane and sevoflurane are preferred inhalational agents, while ketamine combined with a sedative like medetomidine is commonly used for injectable anesthesia. The administration technique must be precise to achieve the desired anesthetic depth.

Complication management: Common complications include hypothermia, respiratory depression, and cardiovascular instability. Monitoring and supportive care are essential to promptly address any issues that arise during anesthesia.

### 2.4 Other small mammals (hamsters, gerbils, etc.)

Other small mammals, such as hamsters and gerbils, have species-specific anesthetic needs that require specialized techniques and consideration.

Species-specific anesthetic needs: Understanding the unique physiological and anatomical characteristics of each species is essential for selecting appropriate anesthetic agents and techniques. Factors such as metabolic rate, respiratory function, and stress response must be considered.

Practical techniques: Techniques for inducing and maintaining anesthesia in these species may involve the use of inhalational agents like isoflurane or injectable combinations tailored to the specific animal. Ensuring a smooth and stress-free induction is critical for effective anesthesia.

Monitoring and safety: Continuous monitoring of vital signs and maintaining a stable anesthetic depth are crucial for the safety and well-being of these animals.

Postoperative care focuses on minimizing stress, preventing hypothermia, and ensuring a smooth recovery.

## 3. Anesthesia in large laboratory animals

### 3.1 Non-human primates

Anesthesia in non-human primates (NHPs) involves stringent ethical and legal considerations due to their close phylogenetic relationship to humans. Key aspects include:

Ethical guidelines: Adherence to institutional and national guidelines for the humane treatment of NHPs.

Legal requirements: Compliance with regulations governing the use of NHPs in research, including necessary permits and documentation.

Anesthetic techniques for NHPs need to be carefully planned and executed. This involves:

Preanesthetic preparation: Comprehensive health assessment, fasting, and premedication to reduce stress and anxiety.

Induction: Use of agents such as ketamine, often combined with sedatives like midazolam or dexmedetomidine.

Maintenance: Maintenance with inhalational agents like isoflurane or sevoflurane, with careful monitoring of vital signs.

Postoperative monitoring for NHPs involves:

Close observation: Monitoring for signs of pain, distress, or complications.

Pain management: Providing appropriate analgesia based on the procedure and individual needs.

Supportive care: Ensuring a quiet and comfortable recovery environment with proper hydration and nutrition.

### 3.2 Pigs

Anesthetic protocols for pigs need to account for their unique anatomical and physiological characteristics. Key components include:

Preanesthetic preparation: Thorough health assessment, fasting, and premedication to reduce stress.

Induction: Use of agents like ketamine combined with a sedative such as midazolam or xylazine.

Maintenance: Maintenance with inhalational agents like isoflurane, with careful monitoring to maintain the appropriate anesthetic depth.

Intraoperative monitoring for pigs involves:

Vital signs monitoring: Continuous monitoring of heart rate, respiratory rate, blood pressure, oxygen saturation, and body temperature.

Supportive measures: Providing thermal support to prevent hypothermia, ensuring proper ventilation, and adjusting anesthetic depth as needed.

Postanesthesia care for pigs includes:

Monitoring: Close observation during the recovery period to detect any signs of complications.

Pain management: Administration of analgesics to manage postoperative pain.

Supportive care: Providing a warm, quiet, and comfortable environment to facilitate recovery with proper hydration and nutrition.

This chapter provides a detailed overview of the specific considerations and techniques involved in anesthetizing large laboratory animals, highlighting the importance of individualized care and meticulous monitoring to ensure the well-being of each animal.

## 4. Specialized anesthesia techniques

### 4.1 Inhalational anesthesia

Inhalational anesthesia involves the administration of anesthetic gases through the respiratory system. Proper equipment and setup are crucial for its safe and effective application.

Anesthetic machine: The anesthetic machine is designed to deliver precise concentrations of anesthetic gases mixed with oxygen. Key components include a vaporizer, flowmeter, and breathing circuit.

Vaporizers: Vaporizers are calibrated devices that convert liquid anesthetics into vapors and allow precise control of their concentration.

Breathing circuits: These include non-rebreathing systems for small animals and rebreathing systems for larger animals. Components such as endotracheal tubes, masks, and breathing bags are selected based on the animal's size and species.

Monitoring equipment: Vital for tracking physiological parameters such as heart rate, respiratory rate, oxygen saturation, and end-tidal $CO_2$ levels.

Several inhalational agents are commonly used in laboratory animal anesthesia, each with specific properties and applications:

Isoflurane: It is widely used due to its rapid induction and recovery, minimal metabolism, and relatively low cost. Suitable for a broad range of species.

Sevoflurane: It is known for its low blood-gas solubility, leading to rapid induction and recovery. Often preferred for procedures requiring quick adjustments in anesthetic depth.

Desflurane: This features the fastest onset and recovery among volatile agents but requires a specialized vaporizer due to its low boiling point.

Inhalational anesthesia offers several advantages but also comes with some disadvantages.

Rapid induction and recovery—allows quick adjustments in anesthetic depth and faster recovery times.

Ease of control—precise control over anesthetic depth through vaporizer adjustments.

Minimal metabolism—most agents are excreted unchanged *via* respiration, reducing the metabolic burden on the animal.

Equipment costs—requires specialized and often expensive equipment. Technical expertise—proper setup and monitoring necessitate trained personnel.

Exposure risks—potential occupational exposure to anesthetic gases if proper scavenging systems are not used.

## 4.2 Injectable anesthesia

Injectable anesthesia involves the administration of anesthetic drugs *via* intramuscular, intravenous, or subcutaneous routes. Common agents include:

Ketamine: A dissociative anesthetic providing analgesia and anesthesia, often combined with sedatives like xylazine or medetomidine for balanced effects.

Propofol: A short-acting intravenous anesthetic ideal for induction and maintenance, offering smooth and rapid recovery.

Alfaxalone: A neurosteroid anesthetic suitable for both induction and maintenance, with a favorable safety profile.

Medetomidine/dexmedetomidine: Alpha-2 adrenergic agonists providing sedation, analgesia, and muscle relaxation, often used in combination with other agents.

Anesthetic protocols must be tailored to the specific needs and physiological characteristics of different species.

Rodents: Typically use combinations like ketamine-xylazine or ketamine-medetomidine. Doses must be carefully calculated based on the animal's weight and health status.

Rabbits: Commonly use ketamine combined with medetomidine or midazolam.

Careful monitoring is essential due to their sensitivity to anesthetics.

Dogs and cats: Often use propofol or alfaxalone for induction, followed by maintenance with inhalational agents. Preanesthetic medication includes sedatives and analgesics.

Non-human primates and pigs: Require tailored protocols considering their size and physiological differences, often involving combinations of ketamine with alpha-2 agonists or benzodiazepines.

Safety considerations in injectable anesthesia focus on minimizing risks and managing potential complications.

Accurate dosage: Precise calculation of drug dosages based on species, weight, and health status to avoid overdose or inadequate anesthesia.

Monitoring: Continuous monitoring of vital signs to detect and respond to adverse reactions promptly.

Supportive care: Provision of fluids, oxygen, and thermal support to maintain physiological stability during anesthesia.

Emergency preparedness: Availability of emergency drugs and equipment to manage complications such as respiratory depression, hypotension, or anaphylaxis.

## 4.3 Balanced anesthesia

Balanced anesthesia, also known as multimodal anesthesia, involves the use of multiple anesthetic agents and techniques to achieve the desired anesthetic state while minimizing side effects. The concept focuses on:

Synergistic effects: Combining drugs that act on different pathways to enhance overall anesthetic effects.

Dose reduction: Lowering the doses of individual drugs to reduce the risk of side effects and toxicity.

Comprehensive management: Addressing anesthesia, analgesia, muscle relaxation, and autonomic stability through a combination of agents.

Effective multimodal anesthesia strategies involve:

Preanesthetic medication: Using sedatives, analgesics, and anticholinergics to prepare the animal for anesthesia.

Induction agents: Employing short-acting agents like propofol or alfaxalone for smooth and rapid induction.

Maintenance anesthesia: Utilizing inhalational agents in combination with injectable drugs to maintain anesthesia.

Analgesia: Incorporating opioids, nonsteroidal anti-inflammatory drugs (NSAIDs), and local anesthetics to provide comprehensive pain management.

### 4.4 Case studies

Case studies illustrate the practical application of balanced anesthesia in various scenarios.

#### *4.4.1 Case study 1: Rodent surgery*

Preanesthetic medication: Medetomidine for sedation and analgesia.

Induction: Ketamine for anesthesia.

Maintenance: Isoflurane for maintaining anesthetic depth. Analgesia: Buprenorphine for postoperative pain management.

#### *4.4.2 Case study 2: Canine orthopedic surgery*

Preanesthetic medication: Acepromazine for sedation and analgesia. Induction: Propofol for smooth induction.

Maintenance: Isoflurane for maintaining anesthesia.

Analgesia: Fentanyl for intraoperative pain relief, followed by meloxicam for postoperative analgesia.

#### *4.4.3 Case study 3: Primate research procedure*

Preanesthetic medication: Midazolam for sedation and anxiolysis. Induction: Ketamine for initial anesthesia.

Maintenance: Sevoflurane for maintaining anesthetic depth.

Analgesia: Morphine for pain management, supplemented with local anesthetic blocks.

The examples in Section 4.4 are rather a wish based on 15 years of experience working with laboratory animals based on available drugs. Of course, during the period of major drug delivery disruptions, anesthesia protocols changed due to the list of available medications.

This chapter provides an in-depth exploration of specialized anesthesia techniques, highlighting the importance of tailored approaches, comprehensive monitoring, and the use of multimodal strategies to ensure the safety and efficacy of anesthesia in laboratory animals.

## 5. Special considerations

### 5.1 Pain management

Accurate pain assessment in laboratory animals is critical for providing effective analgesia and ensuring animal welfare. However, assessing pain can be challenging due to species differences, variations in pain expression, and the inability of animals to verbalize their discomfort. Common methods for assessing pain include:

Behavioral indicators: Changes in behavior, such as decreased activity, reluctance to move, abnormal postures, vocalization, and altered grooming habits, can indicate pain. Specific signs vary by species; for instance, rodents may show decreased nest-building or self-grooming, while larger animals may exhibit more overt signs like limping or guarding a painful area.

Physiological indicators: Physiological responses to pain, such as increased heart rate, respiratory rate, blood pressure, and body temperature, can serve as indirect measures of pain. These parameters should be interpreted alongside behavioral observations.

Pain scales: Species-specific pain scales have been developed to quantify pain levels based on observable signs and physiological parameters. Examples include the Mouse Grimace Scale and the Rat Grimace Scale, which assess facial expressions associated with pain.

Analgesic responsiveness: Administering an analgesic and observing changes in behavior or physiological parameters can help confirm the presence of pain. A positive response to analgesia typically indicates that the animal was experiencing pain.

The choice of analgesic agents depends on the species, type, and severity of pain, and potential side effects. Commonly used analgesics in laboratory animals include:

Nonsteroidal anti-inflammatory drugs (NSAIDs): NSAIDs like carprofen, meloxicam, and ketoprofen are commonly used for mild to moderate pain. They provide analgesic, anti-inflammatory, and antipyretic effects by inhibiting cyclooxygenase enzymes. NSAIDs are generally safe but can cause gastrointestinal and renal side effects, particularly with prolonged use.

Opioids: Opioids, such as buprenorphine, morphine, and fentanyl, are potent analgesics used for moderate to severe pain. They act on opioid receptors in the central nervous system to provide pain relief. Opioids can cause sedation, respiratory depression, and constipation, so careful dosing and monitoring are necessary.

Local anesthetics: Local anesthetics like lidocaine and bupivacaine are used for regional anesthesia and pain management at specific surgical sites. They block nerve conduction and provide effective pain relief without systemic effects. However, care must be taken to avoid systemic toxicity, especially in small animals.

Adjuvant analgesics: Other drugs, such as gabapentin, ketamine, and alpha-2 agonists (e.g., dexmedetomidine), can be used as adjuvants to enhance analgesic effects and reduce the required dose of primary analgesics.

Effective postoperative pain management is crucial for animal recovery and welfare. Key considerations include:

Multimodal analgesia: Combining different classes of analgesics can provide synergistic pain relief while minimizing side effects. For example, using an NSAID with an opioid can provide both peripheral and central pain relief.

Timing and duration: Analgesics should be administered preemptively (before the onset of pain) and continued postoperatively based on the expected duration of pain. Regular dosing schedules ensure consistent pain relief.

Monitoring and adjustments: Postoperative pain should be regularly assessed using behavioral and physiological indicators. Analgesic protocols should be adjusted based on the animal's response, considering factors such as the intensity of pain, species, and individual variation.

Non-pharmacological measures: Environmental modifications, such as providing a comfortable and quiet recovery area, minimizing handling, and offering soft bedding, can help reduce stress and discomfort. Additionally, providing appropriate nutrition and hydration supports overall recovery.

### 5.2 Anesthesia in neonates and juveniles

Neonates and juvenile animals present unique challenges for anesthesia due to their physiological immaturity and rapid developmental changes. Key challenges include:

Immature organ systems: Neonates have immature hepatic and renal systems, affecting drug metabolism and excretion. This immaturity can lead to prolonged drug effects and increased sensitivity to anesthetics.

High metabolic rate: Young animals have a higher metabolic rate, leading to increased oxygen consumption and heat production. This requires careful monitoring of ventilation and body temperature during anesthesia.

Cardiovascular instability: The cardiovascular system of neonates is less able to compensate for changes in blood pressure and volume. This can result in rapid onset of hypotension and bradycardia during anesthesia.

Airway management: Smaller airway size and increased airway resistance make airway management more challenging in neonates. Endotracheal intubation can be technically demanding, and the risk of airway obstruction is higher.

To accommodate the unique physiology of neonates and juveniles, the following specific adjustments to anesthetic protocols are necessary.

Anesthetic dose reduction: Lower doses of anesthetics are required due to immature metabolic pathways and increased sensitivity. Titrating the dose based on the animal's response is essential to avoid overdose.

Selection of anesthetic agents: Agents with a short duration of action and minimal systemic effects are preferred. Inhalational anesthetics, such as isoflurane and sevoflurane, allow rapid adjustment of anesthetic depth and quick recovery.

Temperature regulation: Maintaining normothermia is critical, as neonates are prone to hypothermia. Pre-warming, using heating pads or warm blankets, and monitoring core temperature are essential measures.

Fluid management: Careful fluid administration is necessary to maintain hydration and cardiovascular stability. Isotonic fluids are generally preferred, and the rate should be adjusted based on the animal's size and physiological status.

Ensuring the safety of neonates and juveniles during anesthesia involves Close monitoring: Continuous monitoring of vital signs, including heart rate, respiratory rate, oxygen saturation, and body temperature, is crucial. This allows early detection of complications and timely interventions.

Adequate recovery support: Postoperative care should include a warm, quiet environment with minimal handling. Adequate analgesia and nutritional support are also essential for recovery.

Specialized equipment: Use of appropriately sized equipment, such as small endotracheal tubes, catheters, and monitoring devices, is necessary to accommodate the smaller size and unique anatomy of neonates.

## 5.3 Anesthesia in geriatric animals

Geriatric animals have specific physiological and pharmacological considerations that require tailored anesthetic management.

Decreased organ function: Aging is associated with reduced hepatic and renal function, leading to altered drug metabolism and excretion. This can result in prolonged drug effects and an increased risk of toxicity.

Cardiovascular changes: Older animals often have decreased cardiac output, reduced blood vessel elasticity, and an increased risk of cardiovascular disease. These changes can affect the animal's response to anesthetic agents and increase the risk of hypotension and arrhythmias.

Respiratory compromise: Age-related changes in the respiratory system, such as decreased lung elasticity and reduced respiratory muscle strength, can impair ventilation and gas exchange. This increases the risk of hypoventilation and hypoxia during anesthesia.

Increased sensitivity to anesthetic agents: Geriatric animals may be more sensitive to anesthetics and sedatives, requiring lower doses to achieve the desired effect.

Anesthetic protocols for geriatric animals should be customized to address their specific needs.

Dose adjustment: Lower doses of anesthetic and sedative agents are typically required due to decreased metabolism and increased sensitivity. Anesthetic depth should be carefully titrated to avoid excessive sedation.

Choice of anesthetic agents: Agents with minimal cardiovascular and respiratory effects, such as inhalational anesthetics (isoflurane and sevoflurane) or short-acting intravenous agents (propofol and alfaxalone), are preferred. The use of analgesics with minimal side effects, such as NSAIDs and opioids, should be carefully considered.

Preoperative assessment: A thorough preoperative assessment, including a review of the animal's medical history, physical examination, and laboratory tests (e.g., blood work and ECG), is essential to identify any preexisting conditions that may impact anesthesia.

Effective monitoring and recovery are critical for the safety and well-being of geriatric animals.

Enhanced monitoring: Continuous monitoring of vital signs, including ECG, blood pressure, oxygen saturation, and temperature, is essential to detect and manage any complications promptly.

Postoperative care: Geriatric animals may have a slower recovery from anesthesia due to decreased metabolic rate and impaired organ function. Providing a warm, quiet environment with minimal stress is crucial for a smooth recovery.

Pain management: Adequate pain management is vital, as older animals may have preexisting chronic pain conditions. Analgesic protocols should be adjusted based on the animal's response, considering potential side effects and interactions with other medications.

Nutritional and hydration support: Ensuring adequate nutrition and hydration is important, as older animals may have reduced appetite and fluid intake. Intravenous or subcutaneous fluids may be necessary in some cases.

In summary, the anesthesia of small and large laboratory animals, particularly in special populations such as neonates, juveniles, and geriatric animals, requires careful consideration of their unique physiological and pharmacological needs. By tailoring anesthetic protocols and providing comprehensive monitoring and supportive care, researchers and veterinarians can ensure the safety and welfare of these animals during experimental procedures.

## 6. Conclusion

In animal research, the selection of appropriate anesthetic protocols plays a critical role in ensuring animal welfare and the accuracy of experimental data. However, there are circumstances in which humane euthanasia becomes necessary, either due to the experimental design, ethical concerns, or to alleviate severe suffering. Understanding the physiological effects of anesthesia on animals and recognizing when euthanasia is the most humane option are both essential components of responsible research practices [16, 19].

### 6.1 Importance of selecting the right anesthetic protocol

Anesthesia impacts several physiological systems, including the respiratory, cardiovascular, metabolic, and immune systems. Researchers must carefully select anesthetic agents, considering factors such as species, the nature of the experiment, and the potential impact of anesthesia on the data. However, despite best practices, complications may arise during procedures that necessitate the need for euthanasia to prevent undue suffering.

### 6.2 Physiological impact of anesthesia

Anesthesia causes changes in various systems.

- Respiratory depression: A common side effect, especially with inhalational agents, that can result in hypoxia.

- Cardiovascular effects: Hypotension and arrhythmias that may compromise the well-being of the animal, especially if complications arise during or after surgery.

- Metabolic and neurological effects: Metabolic suppression, hypothermia, and CNS depression may result in protracted recovery or irreversible damage.

If an animal experiences severe, unresolvable physiological instability due to anesthesia or the research protocol itself places the animal under conditions of severe stress, and euthanasia may be ethically required.

### 6.3 Implications for experimental data

Anesthesia can confound data in studies focused on respiratory, cardiovascular, or metabolic function. In cases where anesthesia-induced complications compromise the integrity of the experiment or the animal's health, euthanasia may be the most humane choice to end suffering and avoid collecting unreliable data.

### 6.4 Cases for euthanasia

Euthanasia is required in the following situations:

- Irreversible anesthetic complications: If an animal experiences severe respiratory depression, cardiac arrest, or unmanageable hypotension during anesthesia, euthanasia is often necessary to prevent prolonged suffering and ensure ethical compliance.

- Inadequate recovery: Prolonged postanesthetic recovery, characterized by persistent hypothermia, immobility, or pain that cannot be adequately managed, may warrant euthanasia to avoid further distress.

- Experimental endpoints: In certain studies, animals are anesthetized for terminal procedures, such as tissue collection or experimental surgery, where euthanasia is an intended part of the protocol to ensure humane treatment. These procedures are designed to minimize suffering while obtaining critical data.

- Unrelievable pain or distress: If anesthesia fails to control pain or if severe complications arise postoperatively, such as organ failure, euthanasia may be required according to ethical guidelines. This is particularly relevant in cases where the animal's suffering cannot be alleviated through medical intervention.

### 6.5 Guidelines for humane euthanasia

Euthanasia must be carried out according to institutional guidelines, such as those provided by the American Veterinary Medical Association (AVMA) or other ethical bodies. Common euthanasia methods include:

- Overdose of anesthetic agents: Isoflurane or barbiturate overdose is often used for euthanizing animals, as it induces a rapid and painless death.

- Physical methods: In small animals, physical methods like cervical dislocation or decapitation may be performed under anesthesia to ensure humane treatment.

### 6.6 Optimizing anesthetic protocols

To avoid unnecessary euthanasia, researchers must:

- Monitor physiological parameters: Continuous monitoring of heart rate, respiratory rate, and temperature is essential to detect complications early and adjust anesthesia as needed.

- Tailor anesthetic protocols: Individualize anesthetic regimens to minimize the risk of complications based on the animal's species and experimental design.

- Use appropriate controls: Ensure the experimental design includes controls for anesthesia-induced physiological changes to mitigate their effects on data.

In conclusion, the administration of anesthesia plays a critical role in ensuring humane treatment in animal research, but it also significantly impacts physiological processes, which can alter experimental outcomes. The careful selection of anesthetic agents, vigilant monitoring, and appropriate recovery care are essential for minimizing these effects. However, when complications arise or when an animal's suffering cannot be alleviated, euthanasia becomes the most humane option. Understanding these scenarios and adhering to ethical guidelines is crucial for maintaining animal welfare and the integrity of scientific research.

## Acknowledgements

The author acknowledges the use of DeepSeek V3 for language polishing of the manuscript.

## Conflict of interest

The authors declare no conflict of interest.

## Appendix A. Anesthetic drug formulary

### Dosages and administration routes

This section provides a comprehensive list of commonly used anesthetic and analgesic drugs in laboratory animals, including recommended dosages and administration routes. It is essential to adjust these dosages based on the species, individual animal characteristics, and specific research needs.

1. Inhalational anesthetics

   Isoflurane: 1–3% via inhalation for induction and maintenance; suitable for most species, including rodents, rabbits, and larger animals.
   Sevoflurane: \–4% via inhalation for induction and maintenance; preferred for its rapid onset and recovery, commonly used in rodents and non-human primates.

2. Injectable anesthetics

   Ketamine: 10–\0 mg/kg intramuscular (IM) or intraperitoneal (IP) for induction; often combined with a sedative (e.g., xylazine or medetomidine) for enhanced sedation and analgesia.
   Propofol: \–6 mg/kg intravenous (IV) for induction; used primarily in larger animals, such as dogs and pigs, due to the requirement for IV access.

3. Sedatives and tranquilizers

   Xylazine: 1–5 mg/kg IM or IP for sedation; commonly used in combination with ketamine in rodents and larger animals.
   Dexmedetomidine: 0.1–0.5 mg/kg IM or IV for sedation; offers better sedation and analgesia than xylazine with fewer side effects.

4. Analgesics

   Buprenorphine: 0.01–0.05 mg/kg subcutaneous (SC) or IM for moderate pain relief; long-acting opioid suitable for rodents and larger animals.

Carprofen: \–4 mg/kg SC or oral for mild to moderate pain relief; NSAID commonly used in rodents, rabbits, and larger species.

### Species-specific recommendations

Different species exhibit varying responses to anesthetic agents due to differences in physiology, metabolism, and drug sensitivity. Below are some species-specific considerations:

Rodents (mice and rats): Rodents have a high metabolic rate, necessitating careful dose adjustments and frequent monitoring. Isoflurane is preferred for its rapid induction and recovery, while injectable combinations like ketamine/xylazine provide good anesthesia and analgesia.

Rabbits: Rabbits are sensitive to respiratory depression and stress-induced complications. Isoflurane is commonly used, and careful monitoring of ventilation is required. Buprenorphine and meloxicam are commonly used for analgesia.

Non-human primates: Primates require careful preanesthetic assessment and monitoring due to their physiological similarities to humans. Ketamine/dexmedetomidine is a common injectable protocol, while isoflurane provides reliable maintenance.

Canines and felines: Dogs and cats have well-established anesthetic protocols, with propofol or alfaxalone for induction and isoflurane for maintenance. NSAIDs and opioids are commonly used for postoperative analgesia.

Pigs: Pigs are sensitive to stress and require careful handling. Inhalational anesthetics like isoflurane are preferred, and adequate analgesia with NSAIDs or opioids is crucial.

## Appendix B: Case studies

### Practical applications

Case studies provide practical insights into the application of anesthesia and analgesia protocols in various research settings. These examples highlight challenges encountered and solutions implemented:

Case study 1: Anesthesia in rodent neurosurgery: This study describes the use of ketamine/xylazine anesthesia in a rat model for stereotaxic brain surgery. Key challenges included maintaining stable anesthesia and minimizing respiratory depression. The solution involved titrating anesthetic doses and using a heating pad to prevent hypothermia.

Case study 2: Pain management in the postoperative rabbit model: A case involving postoperative pain management in rabbits undergoing orthopedic surgery. The study explored the efficacy of multimodal analgesia combining buprenorphine and meloxicam. Observations included improved pain relief and faster recovery times.

Case study 3: Anesthetic management in non-human primates. This case details the anesthetic management of non-human primates during a cardiovascular study. The challenges included ensuring cardiovascular stability and minimizing stress. The use of dexmedetomidine for sedation and isoflurane for maintenance provided a balanced approach.

In conclusion, these appendixes provide essential information and resources for the safe and ethical administration of anesthesia and analgesia in laboratory animals. By adhering to these guidelines and continuously learning from practical experiences, researchers can uphold high standards of animal welfare and scientific integrity.

## Author details

Petr Peretyagin* and Anna Soloveva
National Research Lobachevsky State University of Nizhny Novgorod, Nizhny Novgorod, Russia

*Address all correspondence to: peretyaginpv@gmail.com

## References

[1] Flecknell PA. Laboratory Animal Anaesthesia. 3rd ed. Amsterdam, Netherlands: Academic Press; 2009

[2] National Research Council. Guide for the care and use of laboratory animals. 8th ed. Washington, DC: National Academies Press; 2011

[3] Ingeborg L, Niemi SM, Kohn L. Pain management and humane endpoints in laboratory animals. Institute for Laboratory Animal Research (ILAR). 2015;**56**(2):189-195

[4] Smith JA, Boyd KM. Ethical and Welfare Considerations in Laboratory Animal Use. London, UK: Royal Society of Medicine Press; 2009

[5] United States. Animal Welfare Act [Internet]. Washington, DC: U.S. Government; [cited 01.08.2024.]. Available from: https://www.nal.usda.gov/animal-health-and-welfare/animal-welfare-act

[6] Beauchamp TL, Childress JF. Principles of Biomedical Ethics. 8th ed. New York, NY, USA: Oxford University Press; 2019

[7] Rupreht J, van Lieburg MJ, Lee JA. History of Anaesthesia and Pain. Dordrecht, Netherlands: Springer; 2002

[8] Grimm KA, Lamont LA, Tranquilli WJ, Greene SA, Robertson SA. Veterinary Anesthesia and Analgesia: The Fifth Edition of Lumb and Jones. Hoboken, NJ, USA: Wiley-Blackwell; 2015

[9] Thurmon JC, Tranquilli WJ, Benson GJ. The History of Anesthesia in Veterinary Medicine. Lumb & Jones Veterinary Anesthesia. Baltimore, MD, USA: Williams & Wilkins; 1996

[10] Franks NP. General anesthesia: From molecular targets to neuronal pathways of sleep and arousal. Nature Reviews Neuroscience. 2008;**9**(5):370-386

[11] Hemmings HC, Egan TD. Pharmacology and Physiology for Anesthesia: Foundations and Clinical Application. 3rd ed. Philadelphia, PA, USA: Elsevier; 2019

[12] Fish RE, Brown MJ, Danneman PJ, Karas AZ. Anesthesia and Analgesia in Laboratory Animals. 2nd ed. Amsterdam, Netherlands: Academic Press; 2008

[13] Riviere JE, Papich MG. Veterinary Pharmacology and Therapeutics. Hoboken, NJ, USA: Wiley-Blackwell; 2018

[14] Fish RE, Brown MJ, Danneman PJ. Anesthesia and Analgesia in Laboratory Animals. Amsterdam, Netherlands: Academic Press; 2008

[15] Duke T. Monitoring anesthesia in veterinary patients. Veterinary Clinics of North America: Small Animal Practice. 2006;**36**(5):1087-1107

[16] Blake C, Pellett C. The Veterinary Nurse—Anaesthesia in Small Rodents. 2022. Available from: https://www.theveterinarynurse.com/content/clinical/anaesthesia-in-small-rodents

[17] Masutomi N, Shibutani M. Analgesia, anesthesia, and postoperative care in laboratory animals. In: Tatlisumak T, Fisher M, editors. Handbook of Experimental Neurology: Methods and Techniques in Animal Research. Cambridge, UK: Cambridge University Press; 2006. pp. 40-66

[18] Navarro KL, Huss M, Smith JC, Sharp P, Marx JO, Pacharinsak C. Mouse

anesthesia: The art and science. ILAR Journal. 2021;**62**(1-2):238-273. DOI: 10.1093/ilar/ilab016

[19] Hawkins M, Pascoe P. In: In: Ferrets, Rabbits, and Rodents. St., editor. Anesthesia, Analgesia, and Sedation of Small Mammals. Louis, MO, USA: Saunders (an imprint of Elsevier); 2012. pp. 429-451. DOI: 10.1016/B978-1-4160-6621-7.00031-2

Chapter 7

# Histopathological and Behavioral Irreversible Damage Derived from Chronic Exposure to Vanadium Pentoxide Is Similar to that Found in Alzheimer's Disease

*Maria Rosa Avila-Costa, Claudia Dorado-Martínez, Enrique Montiel-Flores, José Luis Ordoñez-Librado, Ana Luisa Gutierréz-Valdez, Cesar Alfonso Garcia Caballero, Javier Sanchez-Betancourt, Leonardo Reynoso-Erazo, Rocío Tron-Alvarez, Vianey Rodríguez-Lara, Jesús Espinosa-Villanueva and Patricia Aley-Medina*

## Abstract

Chronic exposure to vanadium pentoxide ($V_2O_5$) has been linked to neuronal damage equivalent to that found in Alzheimer's disease (AD). Prior research has indicated that chronic inhalation of $V_2O_5$ results in cellular and behavioral changes akin to those observed in AD. A total of 40 male Wistar rats were categorized into two control groups and two experimental groups for the study. The experimental inhaled $V_2O_5$ for 6 months; after 6 months, two groups (control and exposed) were left in a 6-month recovery phase. All groups were evaluated monthly for 6 or 12 months in a spatial memory test. To measure histological alterations, hippocampus CA1, entorhinal and frontal cortices, amygdala, and subiculum underwent Congo red or argentic Bielschowsky impregnation and were analyzed. Memory results show impairment since the third month. The group left in recovery did not show improvement. Chronic $V_2O_5$ inhalation is an AD model that causes irreversible alterations in spatial memory, accumulation of Aβ plaques, accumulation of NFTs, and loss of dendritic spines with no recovery. These alterations are not reversible after 6 months, indicating that the damage increases once the neurodegenerative process is established and the homeostasis is broken. This model characterizes sporadic AD, representing more than 95% of cases.

**Keywords:** Alzheimer's disease experimental model, Aβ plaques, cell death, dendritic spine loss, inhaled exposure, neurofibrillary tangles, vanadium pentoxide



IntechOpen

## 1. Introduction

Alzheimer's disease (AD) is a gradual neurodegenerative disease that provokes cognitive decline with concomitant memory and executive function loss [1], followed by subsequent severe deficits attributed to neuronal and synaptic loss [2]. AD neuropathological hallmarks include amyloid plaques comprised of accumulated amyloid-β (Aβ) peptides, identified as senile plaques, and the neurofibrillary tangles (NFTs) known as the accumulation of hyperphosphorylated tau protein [3, 4]. There is compelling evidence suggesting that while the primary indicator of AD may be heightened Aβ levels, it ultimately leads to cognitive dysfunction and neurodegeneration by triggering alterations in tau. This highlights the urgency of understanding and addressing these underlying mechanisms in pursuing effective treatments [5, 6]. Although significant research has lately been undertaken to consider the impact of both Aβ plaques and tau hyperphosphorylation, Aβ is still considered the triggering element for AD progress. The "amyloid cascade theory" posits that the deposition of plaques serves as the primary event in the onset of AD. Subsequently, factors such as NFTs, neuronal death, damage, and AD-related behavioral outcomes emerge as consequences of this process. AD-associated behavioral outcomes result from this circumstance [6, 7]. Nonetheless, some proof has proposed a more forthright role of tau pathology. For example, the progressive staging of tau pathology [8] is associated better with cognitive decline than Aβ plaques [9, 10].

Experimental models are essential for comprehending AD pathogenesis and implementing novel treatments in preclinical trials. Since 1995, more than 100 AD transgenic mouse models have been created that overexpressed the mutated amyloid precursor protein (APP) or APP/presenilin1 (PS1) cDNA (first and second-generation models) [11]. Without tau pathology, these models effectively form most of the disease's pathological hallmarks, such as neuroinflammation, Aβ plaques, and cell death. The second-generation models, besides forming neuroinflammation, cell death, and Aβ plaques, induce cognitive impairments in an age-dependent manner. However, it cannot be employed for preclinical therapy research due to anti-Aβ antibodies low affinity. The disadvantage of the second-generation models is that the pathology may take more than 18 months to become satisfactorily evident for experimental investigation. Nonetheless, this experimental model effectively modulates Aβ pathology by genome editing to demonstrate the differential functions of neprilysin and insulin-degrading enzymes in Aβ metabolism, recognizing subtypes of somatostatin receptors implicated in Aβ neprilysin-degradation [11]. Afterward, a new double knock-in line carrying the AppNL−F and Psen1P117L/WT mutations was formed, and as a pathogenic consequence, it was synergistic. This third-generation model is distinctive in that it reveals more cored-like Aβ plaque pathology and neuroinflammation than the first and second-generation lines and, therefore, is more appropriate for preclinical analyses of disease-modifying therapies focusing on Aβ plaques.

To date, most tauopathy mouse models have been incapable of repeating the tau pathology without over-expressing mutant human hyperphosphorylated tau protein. As a novel *in vivo* model for analyzing human tauopathy, human MAPT knock-in animals have been found where the whole Mapt gene, all introns and exons, are humanized [12]. The Mapt and MAPT genes are responsible for encoding human and murine tau proteins in both strains, showcasing the remarkable genetic similarities between the two species. This study involved crossing MAPT knock-in mice with single App mice to investigate the role of the Aβ-tau axis in potential AD development. The double knock-in mice exhibited more pronounced tau hyperphosphorylation than the MAPT knock-in mice. However, there was no evidence of tau pathology or neurodegeneration in these mice even when they reached 24 months. In both the absence and presence of Aβ amyloidosis, humanization of tau has significantly accelerated the propagation of AD brain-derived pathological tau [13]. These results indicate that pathological human tau interacts more strongly with human tau than murine tau, suggesting a species-specific preference for pathogenic proteins [11]. As can be seen, to date, the most frequently used models are animal models, almost simply consisting of transgenic mice that overexpress human protein genes associated with familial AD, comprising Aβ plaques (human APP or in combination with human PSEN1) or NFTs (human MAPT) [14–16]. However, AD is characterized by the coexistence of NFTs and Aβ plaques [15]. Other experimental models have incorporated invertebrates such as *C. Elegans* and *Drosophila* and vertebrates such as zebrafish; nevertheless, they are slightly contemplated since these models are very different from human physiology [11, 17, 18].

It is well known that most AD cases are idiopathic, so in this sense, the few sporadic AD animal models are primarily established on the neurotoxins administration *via* stereotaxic, such as cholinergic antagonists, Aβ oligomers, ibotenic acid, and cell cultures [17, 18]. Nevertheless, none characterize the disease's progressive and chronic degenerative characteristics in humans. In summary, some experimental models show only the Aβ deposition that defines AD. This frequently provokes evident memory alterations. However, it is important to note that these models frequently fail to replicate the primary pathological characteristics of AD, including cell death and, notably, the presence of NFTs [15].

On the other hand, most AD cases are idiopathic, and the reasons underlying sporadic AD are uncertain. Old age is frequently acknowledged as the foremost cause of the disease [19]. However, the disease causes are still undetermined [17], so it is essential to develop AD animal models that reproduce the disease better accurately. The rise in neurodegeneration cases in industrialized countries is deeply concerning. This surge is directly linked to the excessive pollutants emitted by internal combustion engines. As we age, the impact of these pollutants becomes increasingly detrimental to our neurological health. Moreover, it is crucial to note that air pollution has been strongly associated with the heightened presence of misfolded and altered proteins, including Aβ and alpha-synuclein. These findings underscore the potential role of air pollution in the progression of specific neurodegenerative conditions [20, 21].

Studies on heavy metal pollution [22] and their consequent biological systems' toxicopathological effects have significantly been documented over the past years [23–25]. In this way, Vanadium (V), a transition metal, corrosion resistant, atomic number 23, is a chemical element classified under the periodic table's Group 5 (VIB), is abundantly available in nature, and prominently present in petroleum products. Its versatility extends to creating robust alloys, vibrant paints, and essential auto parts, making it an indispensable component in various industries [22, 26]. Biologically, V has different effects: essential in trace amounts (0.05 μM) and toxic in excess (>10 μM) [27, 28]. V occurs extensively in the atmosphere, but its abundance is low compared to other elements [22, 29–31]. While V compounds are released into the atmosphere through natural sources like continental dust, volcanos, forest fires, and sea salt sprays [10], anthropogenic activities, mainly fossil fuels burning, industrial activities, and heavy oil discharge, have contributed to environmental V toxic levels [28, 32–35]. The concern regarding V emission to the ambient air begins from the moderately prominent concentrations (20–300 ng/$m^3$) in urban areas, reaching 10 mg/$m^3$ levels in industrialized cities [36, 37]. Despite this, it is crucial to note that approximately 64,000 tons of V are released annually, with a staggering 91%

stemming from coal, crude oil burning, and other mining/metallurgic industries [38]. Additionally, it is worth noting that V is frequently found at elevated concentrations in fossil fuels, primarily in Mexican and Venezuelan petroleum [39]; V exposure poses a significant risk, mainly stemming from gasoline ignition and the inhalation of suspended particulate matter (PM), primarily in the fine aerodynamic range (i.e., PM2.5; <2.5-μm diameter) [34, 38–40].

V toxicity will depend on its valency ($1^+$ to $5^+$) [41], the administration route, and the compound. It has been reported that V has several oxidation states, the most cytotoxic being vanadium pentoxide ($V_2O_5$) [26, 42], dose [43], and duration [25] of exposure. V enters the body parenterally or by ingestion, mainly inhalation [26, 44]. V compounds absorption depends on the exposure route and solubility [26]. Moreover, V compounds solubility and alveolar and mucociliary clearance regulate the absorption rate in the respiratory tract [45]. It has been calculated that about 25% of soluble V is absorbed through the respiratory tract [26, 27, 40, 45]. However, V compounds are scarcely absorbed in the gastrointestinal tract (but well absorbed *via* inhalation). A review of experimental models suggests that the quantity of V absorbed by the gastrointestinal tract is less than 1–2%, where vanadate ($V^{5+}$) is absorbed approximately three times more effectively than vanadyl ($V^{4+}$) [26]. V absorbed through the lungs, gastrointestinal tract, or parenteral routes is transferred in the blood mainly as $V^{4+}$ (since vanadate is reduced by glutathione in erythrocytes), generally bound to albumin and transferrin [46, 47]. Removal of inhaled V is biphasic with a primary fast rate (10–20 h) followed by a more extended phase (40–50 days) [22, 48]. Hence, the respiratory route is the potential occupational and environmental V exposure route. The direct link between the upper respiratory tract and the olfactory bulb, facilitated by olfactory neurons crossing the cribriform plate, offers a compelling pathway for air pollutants infiltrating the brain tissue [49]. Therefore, the brain is a conceivable goal for inhaled V [22].

In this regard, it has been proposed that V acts like phosphorylase due to its structure [26, 44, 50, 51]. Moreover, it has been reported that $V^{5+}$ inhibits tyrosine phosphatase by rising p-Tyr-Ser phosphorylated residues and the expression of the G protein p21 RAS, which is linked to oxidative stress onset [52, 53]. Additionally, it has been described that mice exposed, *via* inhalation to $V_2O_5$ show tubulin polymerization alterations, damaging the structure of the microtubules in the testicular parenchyma and mesenchyme [54]; V phosphorylates tubulin changing its polarity, consequently altering the cytoskeleton structure [55]. Recently, our group demonstrated that rats exposed to $V_2O_5$ (0.02 M) for 6 months showed alterations in spatial memory, probably due to the presence of Aβ plaques, NTFs, loss of dendritic spines and neuronal death in the amygdala, subiculum, hippocampus CA1, and the frontal and entorhinal cortices, showing that neuronal death is similar to that observed in patients who had suffer from AD [56, 57]; with which we established a novel and reliable AD model. It is necessary to analyze whether the alterations in spatial memory and the pathological changes observed in the different brain structures are reversed after the animals have stopped inhaling $V_2O_5$.

## 2. Experimental procedures

A total of 40 male Wistar rats weighing 180–200 g at the start of the study were kept in plastic cages under standard lighting conditions (12-hour light/dark cycle). They were provided with Purina rat chow and water *ad libitum*. The experimental

*DOI: http://dx.doi.org/10.5772/intechopen.1007924*

protocol was conducted in compliance with the Animal Act of 1986 for Scientific Procedures, the Mexican Guideline for Animal Welfare (NOM-062 – ZOO-1999, México), and was approved by the UNAM Ethical Commission (approval number: 1136). We made efforts to minimize the number of animals used and to prevent unnecessary suffering. The rats were approximately two and a half months old at the start of the experiment.

Of the 40 rats, we formed four groups: 10 control and 10 exposed to $V_2O_5$ for 6 months. Of the remaining 20, 10 were control, and 10 were exposed to $V_2O_5$, but they were left in recovery for 6 more months. Thus, after finishing the six inhalation months, the animals were 8 and a half months old, and after 6 months of recovery, they were 14 and a half months old. Therefore, the animals are considered to be middle-aged adults [58].

### 2.1 Spatial memory

Spatial memory was assessed in the T-maze, and training and testing were conducted at 11 AM [59]. Twenty-one days before starting $V_2O_5$ inhalation, we trained the animals. During the first week, the habituation week, a rat was placed at the beginning of the maze and allowed to wander freely for 5 minutes. Each animal repeated this activity five times during the seven-day phase. The acquisition phase took place during the second week. Half of the animals were trained to go to the left side and the other half to the right side; at the end of each arm, left or right, a reward (food pellet) was placed (**Figure 1**), so the animals were deprived of food 12 hours before training or testing, trying to keep 90% of their body weight. Each animal was placed separately in the long arm of the maze, with a gate covering the arm opposite to the training (right or left). This procedure lasted a maximum of 2 minutes, and each rat repeated the test ten times. During the evaluation week (third week), the rats

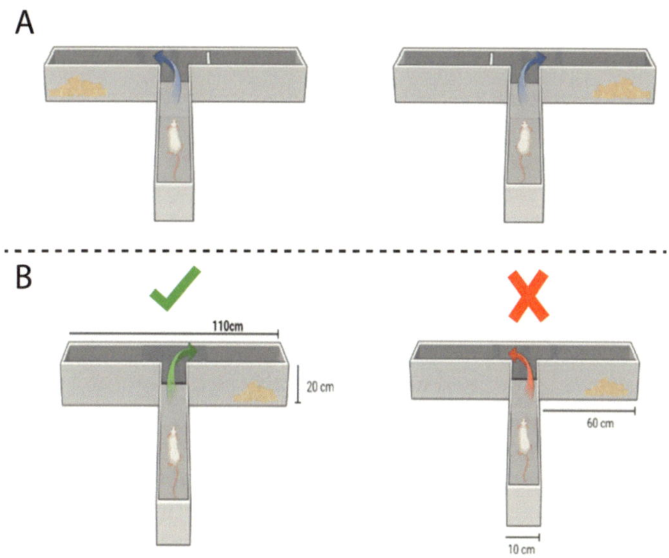


**Figure 1.**
*T-maze experiment procedure: (A) Acquisition phase: 20 rats were trained to turn either right or left at the end of the T-maze. (B) Evaluation phase: If a rat goes toward the arm where it previously found food, it is considered a successful choice. If it goes toward the opposite arm, it is considered a mistake.*

were evaluated according to the arm they had been trained to use. They were placed at the beginning of the maze, and the rat had to reach the right or left side as appropriate and eat the pellet [59].

If the animal went to the correct arm and took the food, it was considered a success; if it did not, it was considered a failure. Each animal was tested five times over the 7 days of the evaluation week and was given 2 minutes to complete the task.

Half of the animals were exposed to the inhalation of $V_2O_5$ after the baseline evaluation. Afterward, spatial memory was tested each month using the above-mentioned procedure. Each time each rat completed the test, the maze was cleansed with 20% alcohol. The spatial memory evaluation was always done in a different room than the one they lived in, always with controlled dim lighting and with the same experimenter.

### 2.2 Vanadium pentoxide inhalation

After the baseline evaluation, the control group was exposed to deionized water inhalation, while the experimental group inhaled 0.02 M $V_2O_5$ [56]. Both groups were exposed for 1 hour three times a week for 6 months. Six months × 4 weeks × 3 inhalations per week = 72 exposures to $V_2O_5$. Inhalations were made in a closed acrylic chamber (40 cm wide × 70 cm long × 25 cm high) connected to an ultra-nebulizer (Shinmed, Taiwan) to nebulize the deionized or $V_2O_5$, keeping a 10 l/min continuous flow. Based on the manufacturer's instructions, about 80% of the particles reaching the animals would be estimated to have a mass mean aerodynamic diameter between 0.5 and 5 mm. After each inhalation, we quantified the V concentrations in the inhalation chamber [57]. The standard V concentration was $1436 \pm 225\ \mu g/m^3$ during the experiment. During inhalations, the animals were observed for respiration depth, rate, and regularity. The chamber was constantly supervised for V level, oxygen concentration, and temperature [56, 57]. At the end of the exposure time, ten rats from the control group (10/20) and ten from the $V_2O_5$-exposed group (10/20) were left to recover. The animals were sacrificed at the end of the behavioral tests and inhalation or recovery periods, and histological techniques were performed.

### 2.3 Euthanasia and perfusion

After six or 12 months, the rats were anesthetized with sodic pentobarbital (lethal dose) and perfused intracardially with 0.9% saline. Subsequently, 10% formaldehyde fixative was administered. The brains were extracted and processed with Bielschowsky and Golgi silver impregnations and Congo red stain.

### 2.4 Histology techniques

The brain tissue was processed using the paraffin technique and then cut on a sliding microtome, obtaining 6 μm thick coronal sections. Using a stereotaxic atlas, we located the frontal and entorhinal cortices, the CA1 hippocampus, the amygdala, and the subiculum to be processed for the Bielschowsky method [60] and the Congo red stain [9, 61]. We meticulously examined each brain structure by analyzing ten slices within a 100 × 100 μm area. Our detailed analysis of digital photomicrographs enabled us to calculate the percentage of damaged pyramidal neurons and neuronal loss for each of the five brain structures. This comprehensive approach provides a deeper understanding of the impact on brain structures and underscores the importance of our findings. The assessment of neuronal loss involved the meticulous counting of all neurons within a 100 × 100 μm area across five distinct brain structures. The subsequent percentage of cell loss was calculated using the following formula:

$$\text{Cell loss\%} = \left(V_2O_5 \text{ exposed cells} - \text{control cells}\right) / \text{control cells} \times 100\% \qquad (1)$$

The tissue was processed using the paraffin technique for the Golgi silver impregnation. On a sliding microtome, 120 μm thick coronal sections were obtained at the dorsal subiculum, hippocampus CA1, frontal and entorhinal cortices, and the basolateral region of the amygdaloid complex. The tissue was processed using the rapid silver Golgi impregnation technique [62]. Dendritic spines were counted as follows: 5 secondary dendrites were selected from 10 pyramidal neurons, and the spines were counted at a length of 10 μm in each structure (see **Table 1**) [63].

| **Amyloid β plaques and Congophilia (cell damage)** | | |
|---|---|---|
| nD | No damage. Neurons with spherical nuclei with a well-preserved nucleus-cytoplasmic relationship and pericentric nucleolus. | 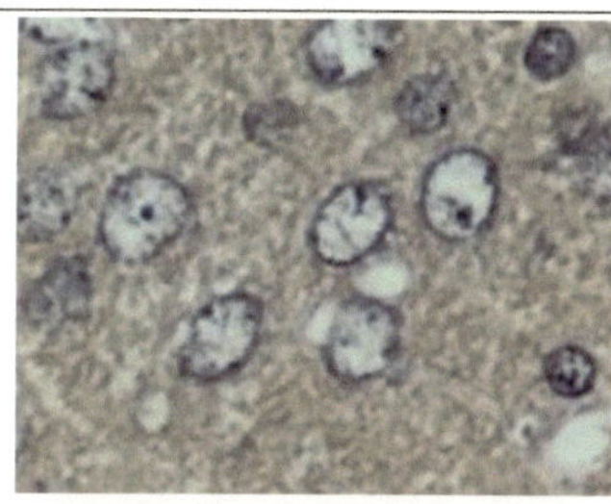 |
| pD | Partial damage. Altered neurons in which the nucleus-cytoplasm ratio is lost, with cytoplasm retraction and congophilic apical dendrite. | 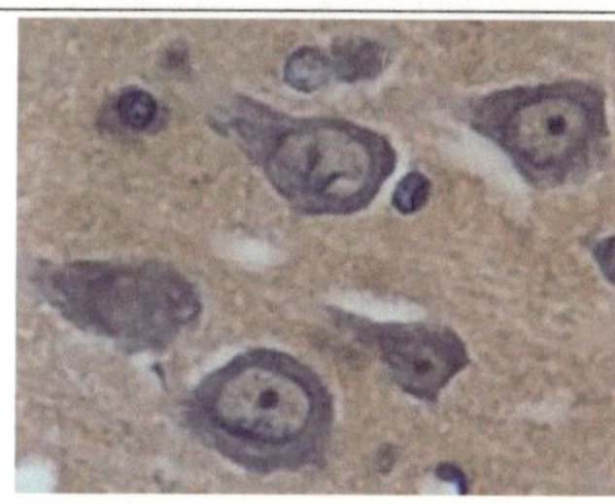 |
| gD | Generalized damage. In neurons that lose nuclei delimitation, the soma appears like an irregular congophilic spot with cytoplasm retraction. | 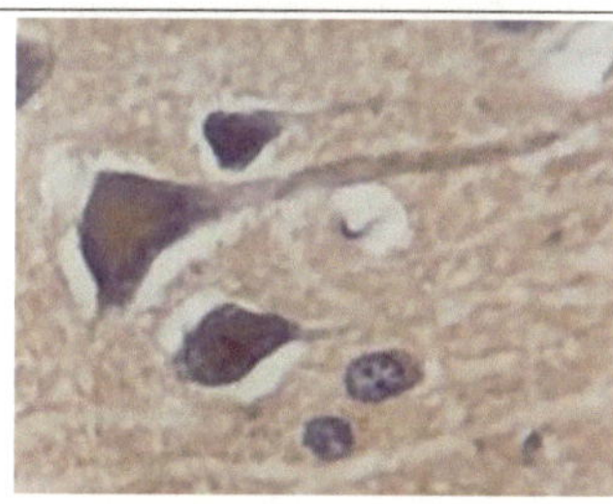 |
| dβA | Diffuse Aβ plaques: extracellular congophilic zones consisting mainly of oligomers and dimers of Aβ. | 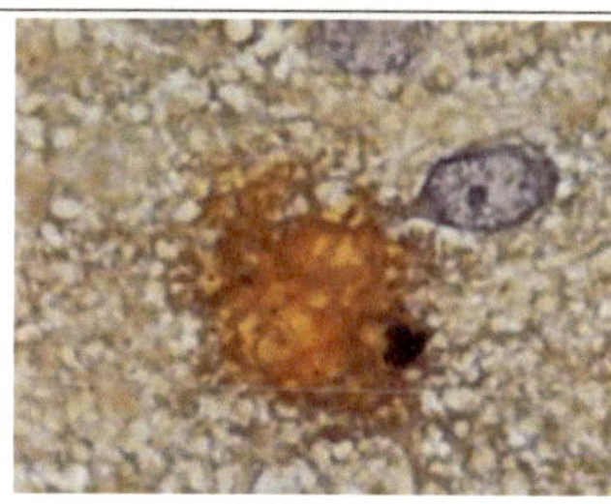 |

| | | |
|---|---|---|
| DcβA | Dense core Aβ plaques: Extracellular congophilic areas comprised of a fibrillar nucleus delimited by a diffuse, non-fibrillar halo. | 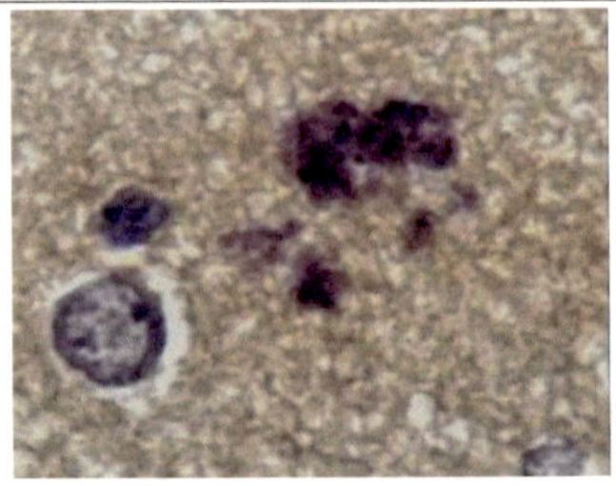 |
| **Neurofibrillary tangles (NFT)** | | |
| preNFT | Pre-tangles, characteristic of healthy neurons, contain visible tau protein but have not yet developed fibers with well-preserved dendrites and concentric nuclei. | 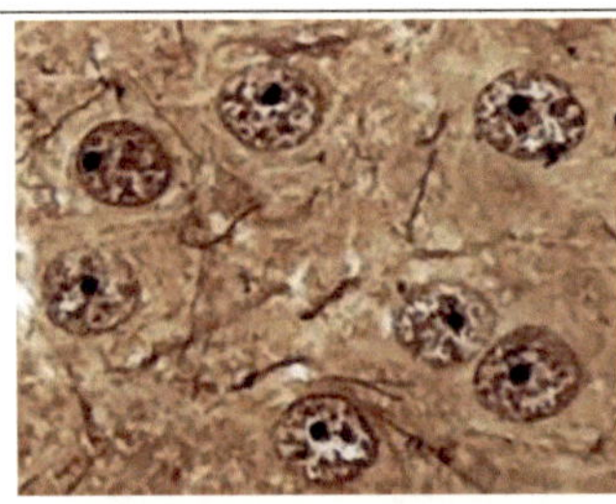 |
| iNFT | The presence of mature or fibrillar intraneuronal NFTs (iNFTs) of hyperphosphorylated tau results in the relocation of the nucleus toward the periphery of the soma, causing notable alterations in dendrite morphology (resembling flames) and the proximal segment of the axon. | 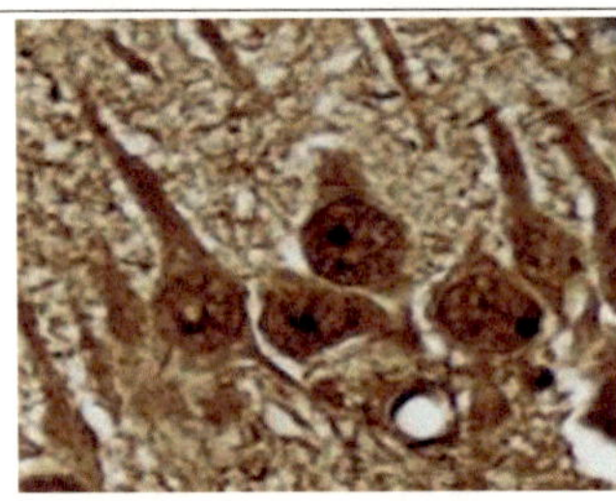 |
| eNFT | External NFTs, known as "Ghost" neurons, emerge as a consequence of neuronal death. They are characterized by the absence of the nucleus and distinct cytoplasmic markings. These eNFTs are detected in the cell soma, displacing the nucleus and compromising cell viability. This leads to the formation of "phantom tangles" or extracellular NFTs, accompanied by clusters of dystrophic neurites. | 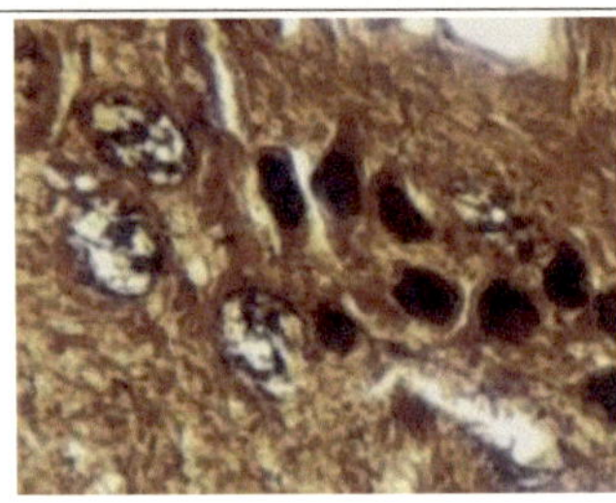 |
| **Dendritic spines** | | |
| | Dendritic spine counting. Five secondary dendrites of ten pyramidal neurons were chosen, and each structure's number of spines within a 10 μm length was counted. | 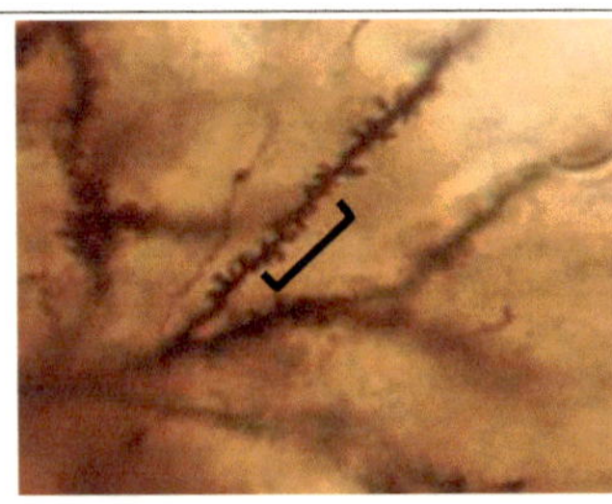 |

**Table 1.**
*Pathology classification of A β Plaques, NFT, and dendritic spines.*

### 2.5 Classification of Aβ plaques, NFTs, and neuronal damage

Two observers blinded to the experimental condition performed quantitative and qualitative histological analysis on images acquired with a Nikon microscope equipped with a Canon EOS Rebel digital camera. The study defined βA pathology, tau NFTs, and neuronal damage as described previously by our group [57], according to what can be seen in **Table 1**. Congo red stain was used to determine Aβ pathology (neuronal damage and Aβ plaques). Neuronal damage (congophilia) was determined between no damage (nD), partial damage (pD), and generalized damage (gD); Aβ plaques: diffuse amyloid-β plaques (dβA) and dense core Aβ plaques (DcβA). Tau NFTs (Bielschowsky silver impregnation) were classified as preNFT (healthy neurons), iNFT (intracellular neurofibrillary tangles), and eNFT (extracellular neurofibrillary tangles) (**Table 1**).

### 2.6 Statistical analysis

Statistical analysis was conducted using GraphPad Prism 10. Data are presented as means ± SEM. Aβ plaques, neuronal damage, dendritic spines, and NFTs were analyzed using one-way ANOVA with the *post hoc* Tukey test. The Kruskal-Walli's test was used to analyze the T-maze results; data were presented as percentages, and *post hoc* comparisons were made with Dunn's test. $P < 0.05$ was considered as statistical significance.

## 3. Results

### 3.1 Spatial memory

After 1 week of training, the T-maze test results (**Figure 2**) show that all the animals achieved at least 80% of successful movements toward the arm of the maze where the reward was located. After 3 months of $V_2O_5$ inhalation, there were statistically significant differences between the groups, with progressive memory deterioration. The cognitive decline is maintained and worsens in the animals left in recovery after $V_2O_5$ inhalation; unlike the animals in the control group, no recovery is observed in the spatial memory deterioration.

### 3.2 Histological analysis

Here, we present the hippocampus CA1, subiculum, entorhinal cortex, amygdala, and frontal cortex Congo red staining (βA plaques and neuronal damage), Bielschowsky impregnation (NFTs), and Golgi stain (dendritic spines) results.

#### *3.2.1 Hippocampus CA1*

According to **Figure 3**, $V_2O_5$ inhalation significantly increased the number of Aβ plaques (**Figure 3B, E**) and decreased the number of healthy neurons (**Figure 3A**) in the hippocampus CA1. These alterations worsened during the recovery time. Likewise,

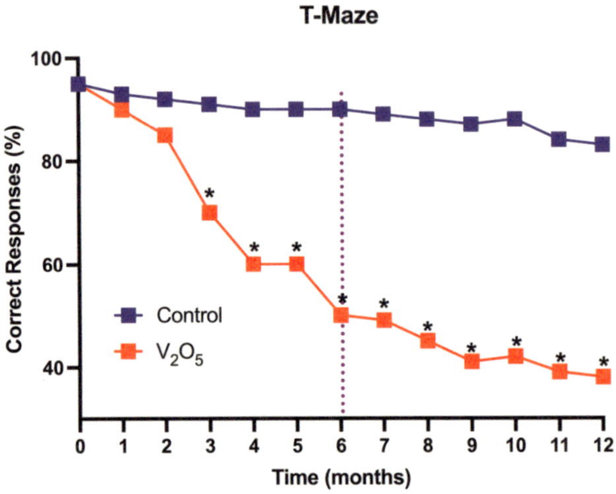


**Figure 2.**
*$V_2O_5$ inhalation effect on spatial memory. The abscissa axis represents the exposure time in months; the ordinate axis shows the correct response percentage. The dotted purple line indicates the time at which $V_2O_5$ exposure was discontinued (Recovery). Kruskal-Walli's test followed by* post hoc *Dunn's test, * = $p < 0.05$.*

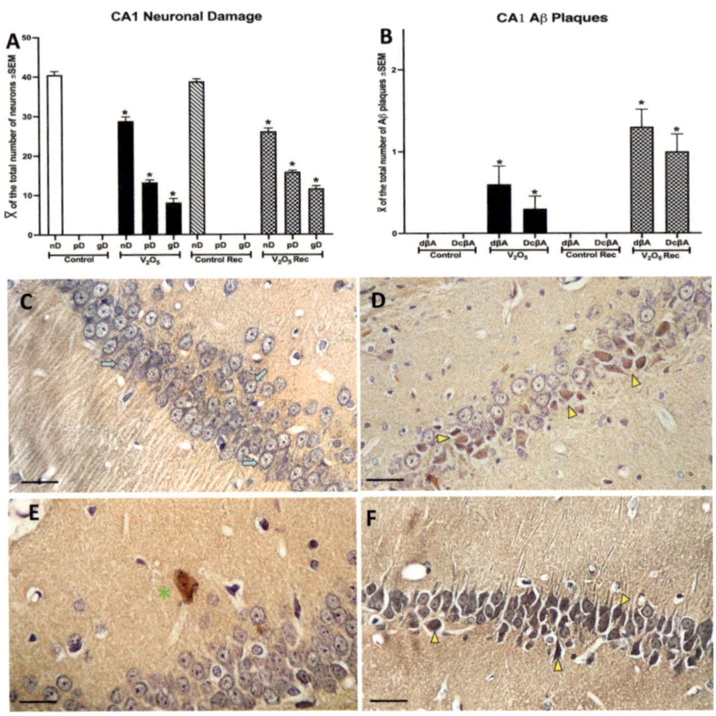


**Figure 3.**
*A. CA1 neuronal damage results; B. CA1 Aβ plaques results, * = $p < 0.05$, ANOVA test followed by Tukey's* post hoc *test. C. Representative Congo-red-stained controls CA1 neurons with the distinctive morphology, the consistently stained nucleus and cytoplasm, and the well-preserved nuclei-cytoplasm ratio (cyan arrows); D. An illustrative instance of $V_2O_5$ CA1 pyramidal neurons stained with Congo red (yellow arrowheads) demonstrates the presence of both generalized and partial damage in certain neurons, characterized by the loss of nucleus boundaries and the emergence of irregularly-shaped congophilic somas; E. Representative $V_2O_5$-exposed CA1 photomicrograph of diffuse Aβ plaque (green asterisk); F. Recovery group representative CA1 pyramidal neurons with generalized damage (yellow arrowheads). Scale bars, 100 μm.*

it can be seen in **Figure 4** that the number of NFTs significantly increased (**Figure 4A**), and the number of dendritic spines decreased compared to control groups (**Figure 4B**); these alterations remained and worsened during the recovery time.

The qualitative results in CA1 from control animals (**Figure 3C**) show characteristic pyramidal neurons with a central nucleolus, a homogeneously stained round nucleus, and cytoplasm around the nucleus with a preserved nucleus-cytoplasm relationship. The $V_2O_5$-exposed animals' neurons (**Figure 3D**) present some cells with a congophilic and amorphous nucleus, loss of the nucleus-cytoplasm relationship, and the apical dendrite is observed to be elongated and highly stained (**Figure 3D**). The CA1 neurons after the recovery period present greater congophilia (damage), they are amorphous, and do not preserve the nucleus-cytoplasm relationship, the apical dendrites are damaged and elongated, and the cytoplasm is retracted (**Figure 3F**).

The average of the total number of CA1 pyramidal neurons decreased by 40% in the $V_2O_5$-exposed group and 62.5% in the recovery group, indicating a significant loss of CA1 pyramidal neurons, which got even worse after the animals stopped inhaling (**Figure 5**).

In Bielschowsky's results for CA1 control animals (**Figure 4C**), the examination reveals healthy pyramidal neurons (preNFT) characterized by a strongly impregnated

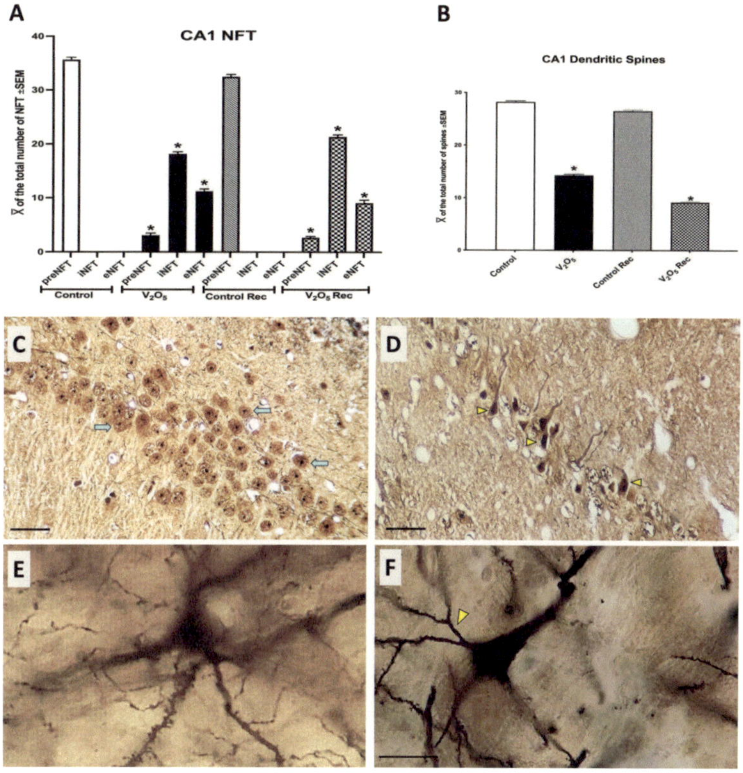


**Figure 4.**
*A. Number of CA1 NFTs; B. CA1 pyramidal neurons Golgi-stain analysis;* $* = p < 0.05$*, ANOVA test followed by Tukey's* post hoc *test; CA1 Bielschowsky's silver impregnation representative photomicrographs. C. CA1 control neurons are without damage with a round nucleus and more impregnated pericentric nucleolus (cyan arrows); D. In the* $V_2O_5$*-exposed and recovery group neurons, it is evident the nucleus and apical dendrite impregnation, the argentophilic neurons show the characteristic flame-shape (yellow arrowheads). E and F show photomicrographs of representative Golgi-stained CA1 pyramidal neurons' dendritic spine density from the control group (E) and* $V_2O_5$*-exposed-recovery group (F). Scale bars, 100 μm (C, D); 25 μm (E, F).*

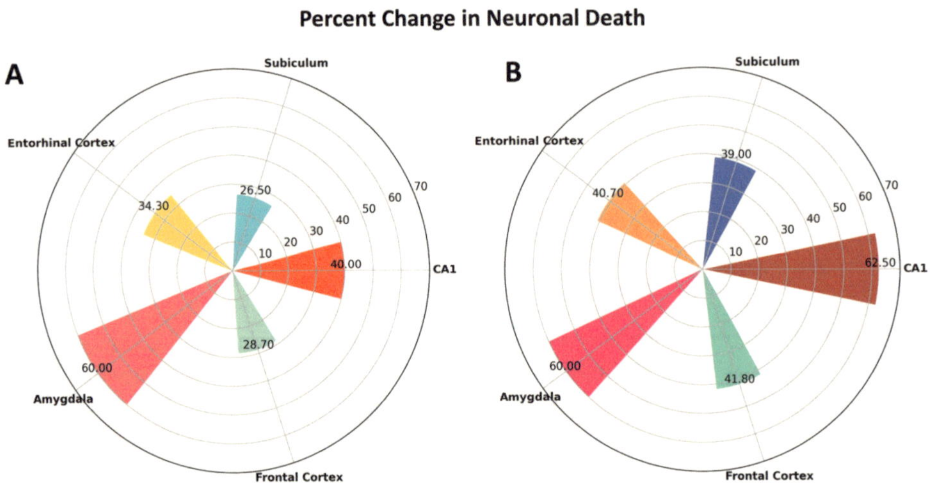


**Figure 5.**
*As can be seen in the figure, the structure that presented the highest percentage of neuronal death was the amygdala, followed by the hippocampus CA1 after $V_2O_5$ inhalation (A) and during the recovery time (B). Although neuronal death was extensive after $V_2O_5$ (A) exposure, it was exacerbated in the structures of the animals left in recovery (B).*

and well-defined central nucleolus. The neurons also exhibit argentophilic round nuclei, with cytoplasm surrounding the weakly impregnated nucleus. The neurons of the $V_2O_5$-exposed animals present abundant neurons with damage, evidenced by the argentophilic nucleus and cytoplasm; the apical dendrite is markedly impregnated (iNFT). After recovery, the neurons appear more impregnated and amorphous, and the cytoplasm (eNFT) is retracted (**Figure 4D**).

As seen in **Figure 4B, E, F**, a significant decrease in dendritic spines was observed after inhalation of $V_2O_5$; this decrease was more evident after the recovery period.

### *3.2.2 Subiculum*

The quantification results of pyramidal neurons in the subiculum Congo red-stained show that nD neurons decreased with a consequent increase in pD and gD neurons after exposure to $V_2O_5$, but these alterations were greater after the recovery period (**Figure 6A**). We observed the same in the number of Aβ plaques (**Figure 6B**), in the recovery group being the most affected.

The total number of subiculum neurons decreased by 26% in the $V_2O_5$-exposed group and 39% in the V-exposed, left-to-recovery group (**Figure 5**).

As shown in **Figure 6C**, the subiculum of control animals shows undamaged pyramidal neurons with homogeneously stained round nuclei and cytoplasm around the nucleus with a well-preserved nucleus-cytoplasm ratio. In contrast, neurons from animals that inhaled $V_2O_5$ present some neurons with perinuclear or cytoplasmic plaques (**Figure 6D, E**). The neurons of the animals after the recovery period show a loss of nuclear-cytoplasmic compartmentalization (**Figure 7D**).

Regarding the results of NTFs in the subiculum (**Figure 7A**), we noticed that the group inhaling $V_2O_5$ and those left in recovery have many tangles, unlike the controls. For the number of dendritic spines (**Figure 7B**), it is evident that $V_2O_5$ produces significant spine loss, worsening after the recovery time.

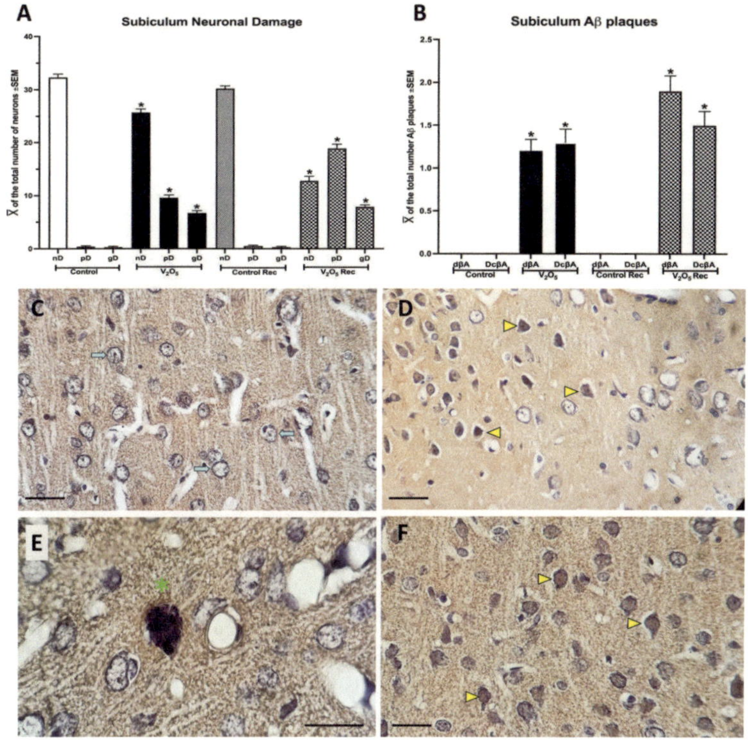

**Figure 6.**
*A. Subiculum neuronal damage results; B. Subiculum Aβ plaques results; * = $p < 0.05$, ANOVA test followed by Tukey's* post hoc *test; C. Representative photomicrographs of control subiculum pyramidal neurons with Congo red-stained. There are evident pyramidal neurons without alterations (cyan arrows); D, Congo-red-stained $V_2O_5$-exposed neurons with partial and generalized damage and loss of compartmentalization (yellow arrowheads); E. Representative $V_2O_5$-exposed-recovery group subiculum photomicrograph of a diffuse Aβ plaque (green asterisk); F. Recovery group subiculum neurons with partial and generalized damage (yellow arrowheads). Scale bar, 100 μm (C, D, F); 50 μm (E).*

With Bielschowsky staining, we observed that the control groups did not show damaged neurons (**Figure 7C**). Unlike the group that inhaled $V_2O_5$ and the recovery group, the neurons present damage, evidenced by the loss of compartmentalization and long argentophilic apical dendrites. After recovery, the neurons appear completely amorphous and have cytoplasmic retraction (**Figure 7D**).

Finally, the dendritic spines are well preserved in the control groups (**Figure 7E**); in contrast, both $V_2O_5$-exposed and recovery groups had significant dendritic spine loss (**Figure 7F**).

### *3.2.3 Entorhinal cortex*

The quantification results of the entorhinal cortex Congo red-stained pyramidal neurons show that more than half of the nD neurons decreased after $V_2O_5$ exposure. This decrease was maintained during the recovery time. The results also show a significant increase in neurons that showed partial (pD) or generalized damage (gD) after $V_2O_5$ exposure, damage that increased after the recovery period (**Figure 8A**). Entorhinal Cortex, cell loss percentage was 34.3% for the $V_2O_5$-exposed group, and

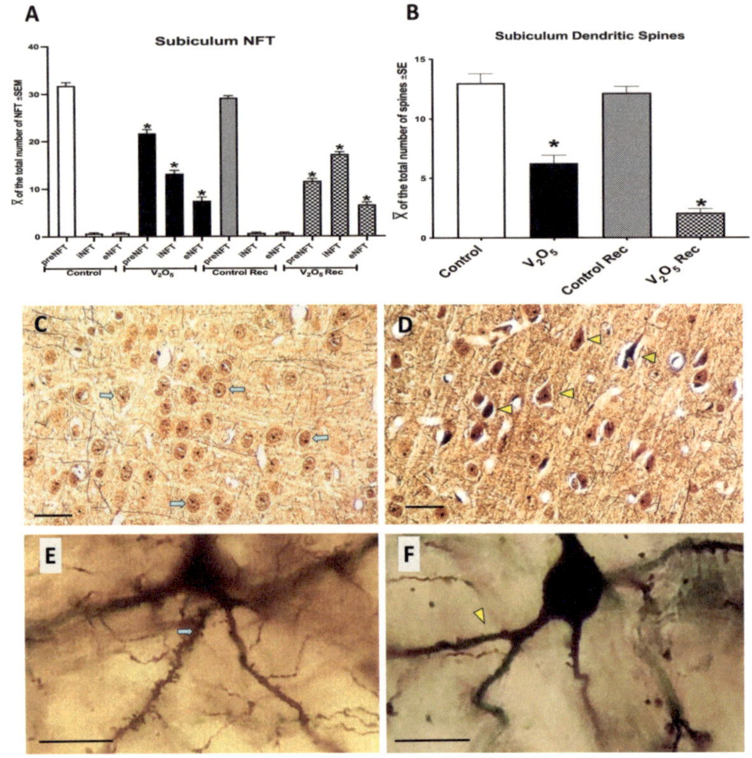


**Figure 7.**
*A. Subiculum NTFs results; B. Subiculum pyramidal neurons dendritic spine count; * = $p < 0.05$, ANOVA test followed by Tukey's* post hoc *test. C-D. Representative Bielschowsky's silver stain photomicrographs of subiculum pyramidal neurons; control group (C) shows characteristic pyramidal neurons without damage (cyan arrows); $V_2O_5$-exposed-recovery group (D) reveals neurons with a long apical dendrite and evident damage with widespread morphological abnormalities (yellow arrowheads). E and F illustrate Golgi-stained subiculum pyramidal neurons' dendritic spine density from the control group (E) and $V_2O_5$-exposed-recovery group (F); $V_2O_5$ inhalation induced a marked decrease in the total number of spines. Scale bars, 100 μm (C, D); 25 μm (E, F).*

40.7% for the recovery group (**Figure 5**). Also, we noted Aβ plaques increase in the experimental groups compared to the controls (**Figure 8B**).

The control group's entorhinal cortex qualitative results (**Figure 8C**) show no damaged pyramidal neurons with a homogeneously stained round nucleus and cytoplasm around the nucleus with a well-preserved nucleus-cytoplasm ratio. The neurons of the $V_2O_5$-exposed animals show alteration of the nucleus-cytoplasm relationship and hyperpigmentation around the nucleus; the neurons of the animals, after the recovery period, show alteration of the nucleus-cytoplasm relationship and cytoplasmic retraction (**Figure 8D**). **Figure 8E** illustrates a Congo red-stained dβA plaque. **Figure 8F** shows a DcβA plaque. These plaques were characteristic mainly in the recovery-$V_2O_5$-exposed group.

With the Bielschowsky silver impregnation, the neurons of the entorhinal cortex of the control groups did not show alterations (preNFT) (**Figure 9A**), unlike the $V_2O_5$-exposed group, where we observed a significant increase in the number of damaged neurons (iNFT + eNFT) In the one maintained during the recovery period after $V_2O_5$ inhalation (**Figure 9A**). **Figure 9C** shows characteristic control groups pyramidal neurons with a round central nucleolus and a well-preserved nucleus-cytoplasm relationship. The neurons of the $V_2O_5$-exposed animals present damaged neurons,

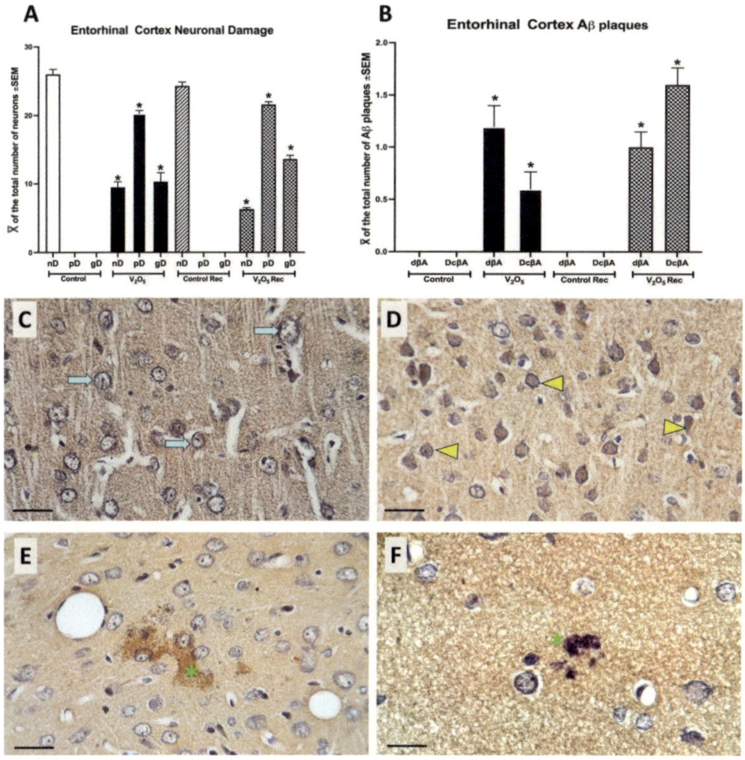


**Figure 8.**
*A. Entorhinal Cortex neuronal damage results; B. Entorhinal Cortex Aβ plaques results; * = $p < 0.05$, ANOVA test followed by Tukey's* post hoc *test; C-F—representative photomicrographs of entorhinal cortex pyramidal neurons Congo red-stained. In C, the control group, there are characteristic pyramidal neurons without alterations (cyan arrows); in D, Congo-red-stained $V_2O_5$-exposed neurons with partial and generalized damage and congophilic cytoplasm (yellow arrowheads); E. Representative recovery-$V_2O_5$-exposed entorhinal cortex photomicrograph of a dβA plaque (green asterisk); F. A characteristic DcβA plaque, seen mainly in the recovery group. Scale bars, 100 μm (C, D, E); 50 μm (F).*

evidenced by a strong impregnation that marks part of the cytoskeleton, including the apical dendrites. After the recovery period, several neurons are amorphous, with a long flame-shaped apical dendrite and cytoplasmic retraction (**Figure 9D**).

When quantifying the density of dendritic spines in the pyramidal neurons of the entorhinal cortex (**Figure 9B**), a decrease was observed in the animals $V_2O_5$-exposed for 6 months, and the spine density did not increase in the group left in recovery (**Figure 9E, F**).

As can be seen in **Figure 9E, F**, the number of dendritic spines displayed by the control groups compared to the $V_2O_5$-exposed groups is significantly different, even after the recovery time.

### *3.2.4 Amygdala*

Our results also demonstrate that animals in the control groups did not have amygdala neuronal damage (**Figure 10A**). However, the groups exposed to $V_2O_5$ presented significant neuronal damage, both partial (pD) and generalized damage (gD), which was more evident in the animals in the group that was left in recovery. The amygdala cell loss percentage was 60% in both experimental groups (**Figure 5**). Qualitative

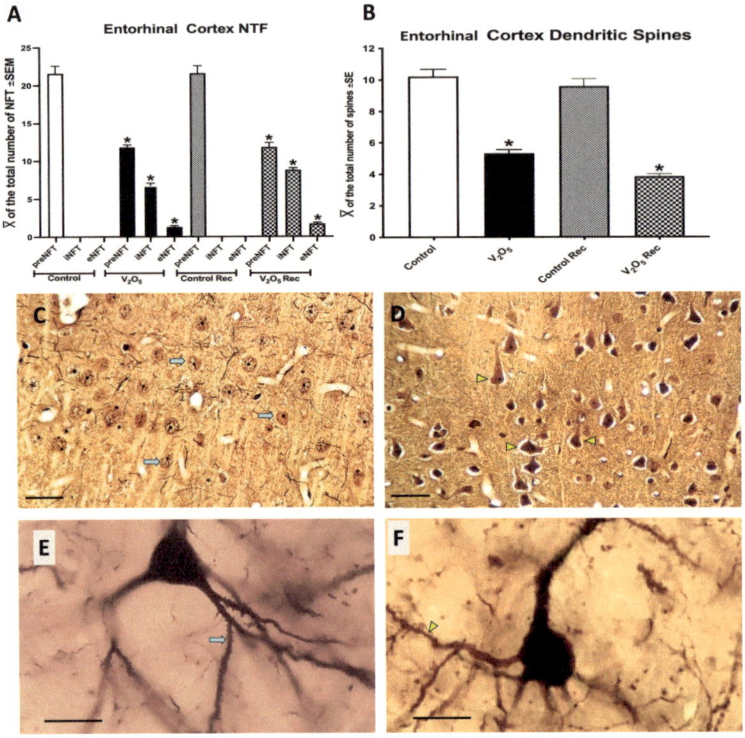


**Figure 9.**
*A. Entorhinal Cortex NTFs results; B. Entorhinal Cortex pyramidal neurons dendritic spine count; * = $p < 0.05$, ANOVA test followed by Tukey's* post hoc *test; C. Representative Bielschowsky's silver stain photomicrographs of entorhinal cortex pyramidal neurons without damage (cyan arrows); in D, recovery-$V_2O_5$-exposed entorhinal cortex reveal neurons with long apical dendrites, flame-shaped and cytoplasm retraction (yellow arrowheads); E, and F, show representative Golgi-stained photomicrographs of entorhinal cortex pyramidal neurons dendritic spine density from the control group (E) and recovery-$V_2O_5$-exposed group (F), $V_2O_5$ inhalation induced a marked decrease in the total number of spines. Scale bars, 100 μm (C, D); 25 μm (E, F).*

results in the amygdala of control animals (**Figure 10C**) show characteristic pyramidal neurons with homogeneously stained round nuclei and cytoplasm around the nucleus with a well-preserved nucleus-cytoplasm ratio. In the $V_2O_5$-exposed animals, the neurons show congophilic cytoplasm, decompartmentalization, and cytoplasm retraction (**Figure 10D**). In the recovery group, neurons with generalized damage and a long, flame-shaped apical dendrite are observed (**Figure 10F**).

**Figure 10B** depicts the quantitative results of Aβ plaques in the amygdala; a significant increase is observed in the number of plaques, both diffuse (dβA) and dense core (DcβA), due to $V_2O_5$ exposure; this increase is more evident after the 6 months of recovery. In the qualitative analysis of the amygdala, we found dβA and DcβA plaques mainly in the recovery group (**Figure 10E**).

With Bielschowsky's silver impregnation in the amygdala, pyramidal neurons show decreased neurons without damage (preNFT) after $V_2O_5$ inhalation and almost 40% during the recovery time. The results show a significant increase in neurons with iNFT accumulation and eNFT after $V_2O_5$ exposure and after recovery (**Figure 11A**). Amygdala qualitative analysis of control animals (**Figure 11C**) shows characteristic pyramidal neurons with impregnated pericentric nucleolus, round nuclei, and cytoplasm around the nucleus with a well-conserved nucleus-cytoplasm ratio. In

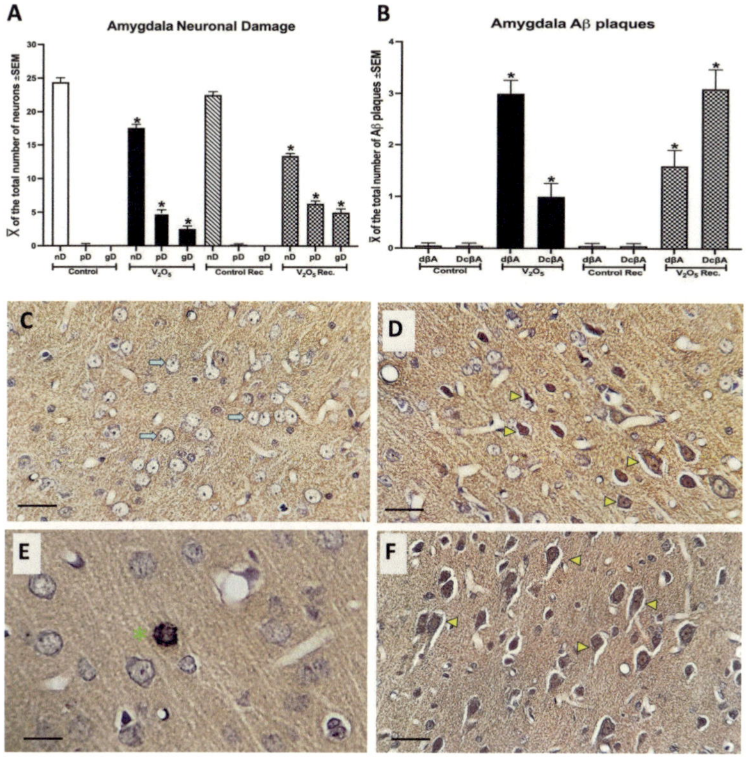


**Figure 10.**
*A. Amygdala neuronal damage results; B. Amygdala Aβ plaques results; * = $p < 0.05$, ANOVA test followed by Tukey's* post hoc *test; C-F. Representative photomicrographs of amygdala pyramidal neurons Congo red-stained. In C, the control group typical neurons are without alterations (cyan arrows); in D, Congo-red-stained $V_2O_5$-exposed damaged neurons with decompartmentalization and cytoplasm retraction (yellow arrowheads); in E. Representative recovery-$V_2O_5$-exposed amygdala photomicrograph of a dβA plaque (green asterisk); F. A representative photomicrograph of amygdala pyramidal cells from the recovery-$V_2O_5$-exposed group where we observed many neurons with partial and generalized damage (yellow arrowheads). Scale bars, 100 μm (C, D, F); 50 μm (E).*

the $V_2O_5$-exposed and recovery animals, neurons show evident cytoskeletal alterations, with loss of compartmentalization and long flame-shaped apical dendrites (**Figure 11D**).

When quantifying pyramidal neurons' dendritic spine density of the amygdala, a decrease was observed in $V_2O_5$-exposed animals, and a greater decrease in the group left in recovery (**Figure 11B, E, F**).

### *3.2.5 Frontal cortex*

Finally, in the frontal cortex, no damage was observed in the pyramidal neurons in any of the control groups (**Figure 12A**); in contrast, in the cells of the $V_2O_5$-exposed animals, a decrease of one-third of the nD neurons was observed after 6 months of inhalation, and almost 80% after the recovery time. Likewise, the results show a significant increase in neurons that showed partial (pD) or generalized damage (gD) after exposure to $V_2O_5$. In the frontal cortex, there was a decrease of (>60%) in the number of neurons after 6 months of exposure and 6 months of recovery (**Figure 12A**). Also, the $V_2O_5$-exposed group showed 28.7% of cell loss and the

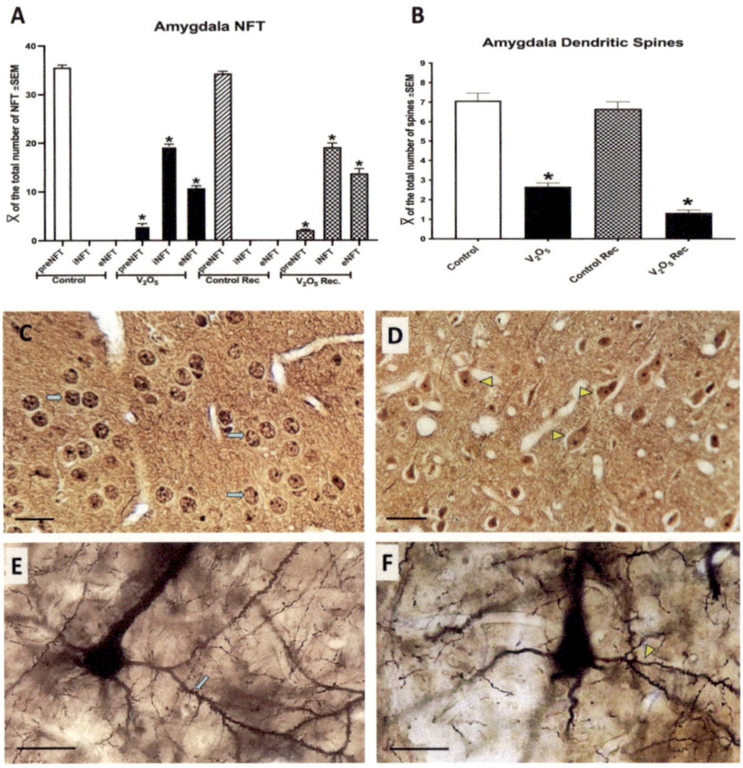


**Figure 11.**
*A. Amygdala NTFs results; B. Amygdala pyramidal neurons dendritic spine count; * = $p < 0.05$, ANOVA test followed by Tukey's* post hoc *test; C. Representative photomicrograph of control amygdala neurons without alterations (cyan arrows); in D, representative Bielschowsky's silver stain photomicrographs of amygdala pyramidal neurons recovery-$V_2O_5$-exposed group showing neurons with long apical dendrites, flame-shaped and cytoplasm retraction (yellow arrowheads). E and F show representative Golgi-stained photomicrographs of amygdala pyramidal neurons' dendritic spine density from the control group (E) and recovery-$V_2O_5$-exposed group (F). Scale bars, 100 μm (C, D); 25 μm (E, F).*

recovery group 41.8% (**Figure 5**). Qualitative results from the frontal cortex of control animals (**Figure 12C**) show pyramidal neurons with homogeneously stained round nuclei, cytoplasm around the nucleus with a well-preserved nucleus-cytoplasm relationship. Neurons from $V_2O_5$-exposed animals present congophilic neurons with nucleus-cytoplasm relationship loss, elongated apical dendrite, and cytoplasm retraction (**Figure 12D**). After the recovery period, the animals' neurons are observed to be amorphous with generalized damage and decompartmentalization (**Figure 12F**).

In the frontal cortex, Aβ plaques increased from an average of 0 plaques per field in the control group to 2.4 dβA and DcβA per field in the $V_2O_5$-exposed group; this process of plaque formation is maintained even after stopping the exposure to $V_2O_5$ to 3.3 plaques per field (**Figure 12B**). In the frontal cortex, diffuse amyloid plaques dβA and dense core amyloid plaques DcβA are observed (**Figure 12E**). In the frontal cortex, the results with the Bielschovsky silver impregnation showed that the controls did not had a decrease in neurons without damage (without deposits of tangles identified as preNFT, compared to the $V_2O_5$-exposed and the recovery groups, where it is observed that these two groups accumulated a significant number of NFTs, both intracellular (iNFT) and extracellular (eNFT), with significant differences (**Figure 13A**). The qualitative frontal cortex results of the control groups (**Figure 13C**), show pyramidal

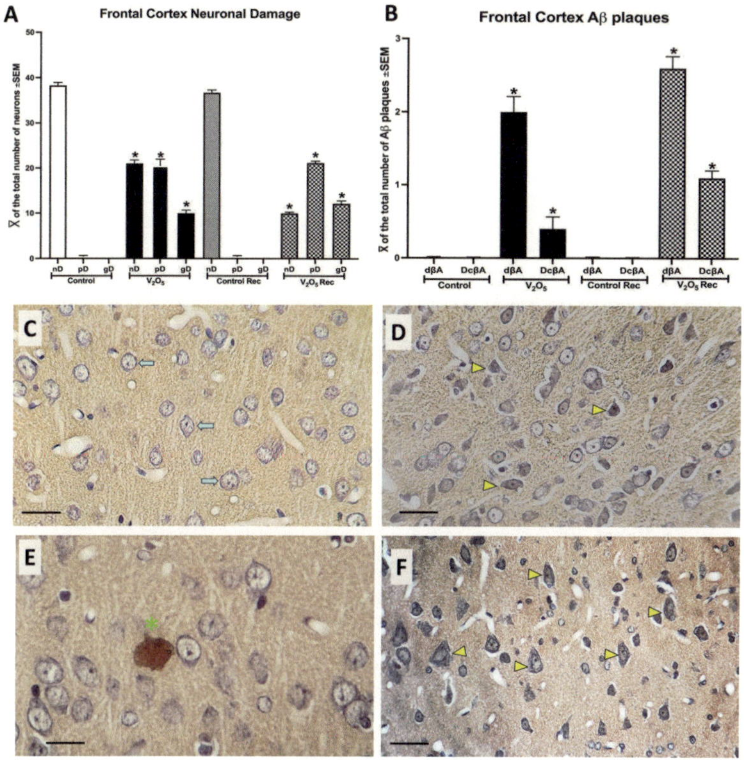


**Figure 12.**
*A. Frontal Cortex neuronal damage results; B. Frontal Cortex Aβ plaques results; * = $p < 0.05$, ANOVA test followed by Tukey's* post hoc *test; D-F—representative photomicrographs of frontal cortex pyramidal neurons Congo red-stained. In C, the control group, there are pyramidal neurons without alterations (cyan arrows); in D, Congo-red-stained $V_2O_5$-exposed damaged congophilic neurons with decompartmentalization, cytoplasm retraction and loss of nuclear-cytoplasmic ratio (yellow arrowheads); E. Representative recovery-$V_2O_5$-exposed frontal cortex photomicrograph of a DcβA plaque (green asterisk). F. Representative recovery-$V_2O_5$-exposed pyramidal congophilic neurons with noticeable damage (yellow arrowheads). Scale bars, 100 μm (C, D, F); 25 μm (E).*

neurons with a central nucleolus, a round nucleus, and a well-preserved nucleus-cytoplasm relationship. The $V_2O_5$-exposed animals had neurons with iNFT, demonstrated by the impregnation that marks the altered cytoskeleton, showing the elongation of the apical dendrite. After the recovery period, some neurons are amorphous, with a very long, flame-shaped apical dendrite and cytoplasmic retraction (**Figure 13D**). Regarding the dendritic spine density, it was observed that the pyramidal neurons of the frontal cortex of the animals in the control groups had the characteristic number of spines in their dendrites; in contrast, both the $V_2O_5$-exposed and the $V_2O_5$-exposed and left in recovery groups, presented a significant decrease in spines compared to the control groups (**Figure 13B, E, F**), being more evident in the recovery group.

As mentioned in the methods section, the percent change in neuronal death was calculated by counting all the neurons in an area of 100 × 100 μm in each of the five brain structures. **Figure 5** shows the percentage of neuronal death observed after $V_2O_5$ inhalation (**Figure 5A**). After recovery (**Figure 5B**), the amygdala and the hippocampus CA1 presented the most significant neuronal loss after inhalation and recovery. Notably, all brain structures analyzed had substantial neuronal loss, mainly after recovery.

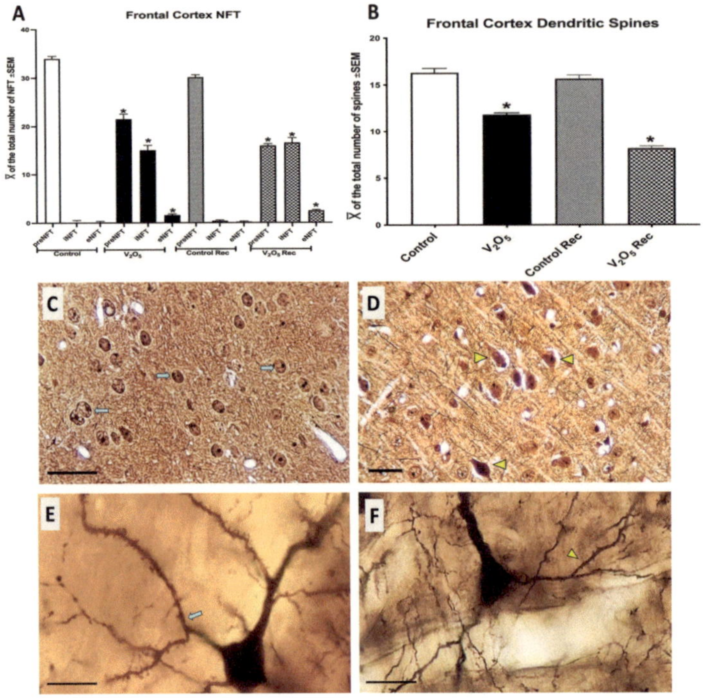


**Figure 13.**
*A. Frontal Cortex NTFs results; B. Frontal Cortex pyramidal neurons dendritic spine count; * = $p < 0.05$, ANOVA test followed by Tukey's* post hoc *test; C. Representative Bielschowsky's silver stain photomicrographs of frontal cortex characteristic pyramidal neurons with no damage (cyan arrows); recovery-$V_2O_5$-exposed (D) reveal neurons with long apical dendrites, flame-shaped and cytoplasm retraction (yellow arrowheads). E and F show representative Golgi-stained photomicrographs of frontal cortex pyramidal neurons' dendritic spine density from the control group (E) and recovery-$V_2O_5$-exposed group (F). Scale bars, 100 μm (C, D); 25 μm (E, F).*

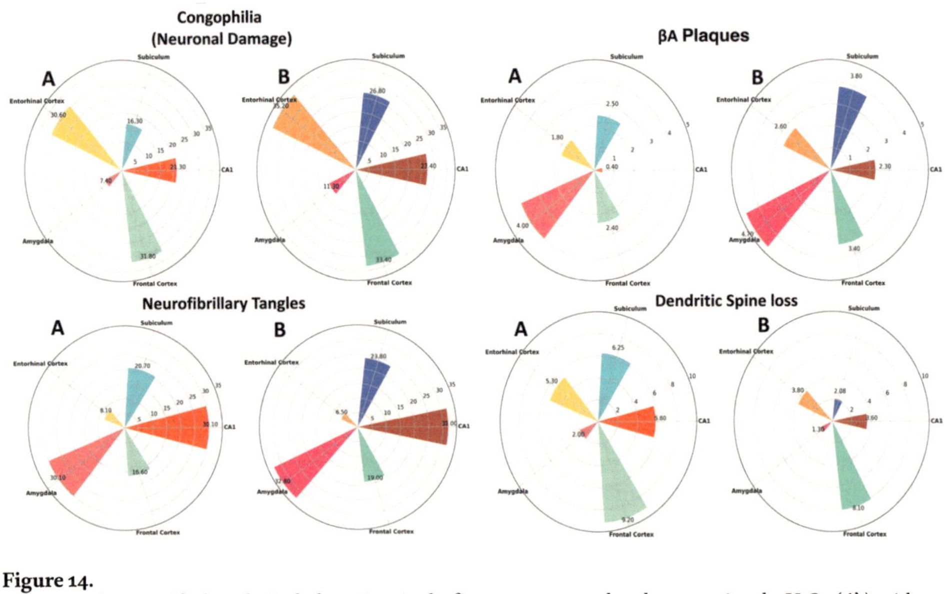


**Figure 14.**
*Compares the quantified cytological alterations in the five structures analyzed, comparing the $V_2O_5$ (A's) with those left 6 months to recover after $V_2O_5$ inhalation (B's).*

*DOI: http://dx.doi.org/10.5772/intechopen.1007924*

**Figure 14** compares the experimental groups' five structures with the data obtained from the cytological analyses. It can be observed that frontal and entorhinal cortices were the brain structures with more neuronal damage in the $V_2O_5$-exposed group (A) and the $V_2O_5$-exposed and left in recovery group (B), followed by the hippocampus CA1. Moreover, the structures that suffered the most significant global damage after $V_2O_5$ exposure (A) and after the recovery period (B) were the frontal cortex, followed by the hippocampus and amygdala. In CA1, the neuronal damage is associated with NFT accumulation and reduced dendritic spine density. In the frontal cortex, neuronal damage is associated with NFTs, Aβ plaques, and loss of dendritic spines. In the subiculum, we found damage mainly related to NFTs and Aβ plaques. We saw damage mainly related to Aβ plaques in the entorhinal cortex and amygdala. It is also evident that the $V_2O_5$-exposed and left-to-recovery group had more significant damage than the six-month inhalation group.

## 4. Discussion

Our data demonstrated significant alterations in all examined structures, compatible with Aβ plaques presence, mainly in the frontal cortex, subiculum, and amygdala; NFTs, predominantly in the hippocampus CA1 and amygdala; synaptic alterations distinguished by the noteworthy dendritic spine loss in all the analyzed structures (**Figure 14**); and consequential cell death (**Figure 5**), where the remaining cells depicted AD-like damage. Furthermore, the group that remained in recovery for 6 months not only did not recover but presented more severe damage than the animals that inhaled V, which means that, once the damage is established, it no longer recovers.

In our AD experimental model generated by the inhalation of $V_2O_5$, behavioral and histological alterations compatible with those observed in AD were observed. The toxic effect of V impacts the brain tissue in several forms. As it has been demonstrated, the most frequent route of exposure to V is inhalation due to suspended particles found in contaminated environments [30, 31, 64], and it is also a nervous system's direct route of entry through the olfactory bulb [49]. The harm happens first in the olfactory mucosa, which triggers the anti-inflammatory and antioxidant systems; when the exposure is chronic, the antioxidant defenses decline and a state of oxidative stress and inflammation appears [65]. V (especially as vanadate) traverses the blood-brain barrier *via* active receptors and metal transporters, provoking the damage related to AD pathology, such as Aβ plaques presence, tau hyperphosphorylation inducing NFTs, oxidative stress, and neuroinflammation inducing the characteristic cognitive dysfunction [22, 53, 66].

We consider it important to explain that airborne V concentrations in the environment are quite variable; in the countryside zones, V concentrations are below 0.001 $\mu g/m^3$. In large cities with raised fossil fuel discharge, the average V airborne concentration varies from 0.03 to 0.4 $\mu g/m^3$ [39]. It has been verified that near industrial areas, V concentrations can range 1 $\mu g/m^3$ [39, 67]. In this work, the V concentration in the inhalation box was 1436 $\mu g/m^3$, surpassing the maximum amounts reported in the atmosphere (1 $\mu g/m^3$). Hence, V concentrations here surpass those described in the metropolises from fossil fuel combustion or occupational exposure [35, 38, 40, 67].

V's neurotoxic properties have been essentially associated with its ability to form reactive oxygen species (ROS) and the consequent oxidative stress that generates the oxidation of phospholipids in cell membranes and, consequently, cell death [68] and neuroinflammation [69]. It can also trigger oxidative stress-related hypomyelination [70] and the reduction of the myelin critical protein [69]. It has also been declared that V provokes apoptotic-like cell death, DNA detachment, and $Fe^{+}$-mediated oxidative stress in neuronal cultures [27] and hippocampal cell death [25, 71]. Likewise, it has been reported that V shuts down protein-tyrosine-phosphatases (PTP) *via* the cysteine catalytic residue binding, which boosts PTP phosphorylation, intensifying the MAPK pathways phosphorylation [27], which feasibly triggers tau protein hyperphosphorylation producing NFTs, as we report in this study.

Consequently, in agreement with our results and the examined publications, V neurotoxicity is analyzed in **Figure 15**: (1) Vanadium arrives at the organism in the tetravalent form ($V^{4+}$ or vanadyl) or as the pentavalent form ($V^{5+}$ or vanadate). It is carried through blood by transferrin and albumin [26, 72]. Vanadyl and vanadate arrive cells throughout anion channels. (2) Then $V^{5+}$ reacts with several antioxidant enzymes like superoxide dismutase (SOD), catalase (CAT), glutathione peroxidase (GSHPx) [73], producing $H_2O_2$ through the Fenton-type reaction. (3) Mitochondria release cytochrome C provoking the apoptotic pathway via caspase 9 activation and consequently caspase 3 [25, 74, 75]. (4) Likewise, $V^{5+}$ produces and spreads ROS concentration, that in turn also promotes the formation of $OH^{\cdot}$, stimulating changes in lipids, proteins, and DNA [27, 68]. (5) $V^{5+}$ is reduced to $V^{4+}$ by NADPH-oxidase to form vanadate again [52, 72]. (6) $V^{4+}$ form is oxidized by $H_2O_2$ to form $V^{5+}$, which will constantly inhibit protein-tyrosine phosphatase (PTP) [69, 76], thus increasing the activity of phosphorylated protein-tyrosine kinase (PTK), (7) starting intracellular signaling pathways [51], (8) causing inflammation via phospholipase-A2 (PLA-A2) and cyclooxygenase 1 and 2 (COX-1/2) activation, stimulating the gliosis process [76, 77]; (9) also generating

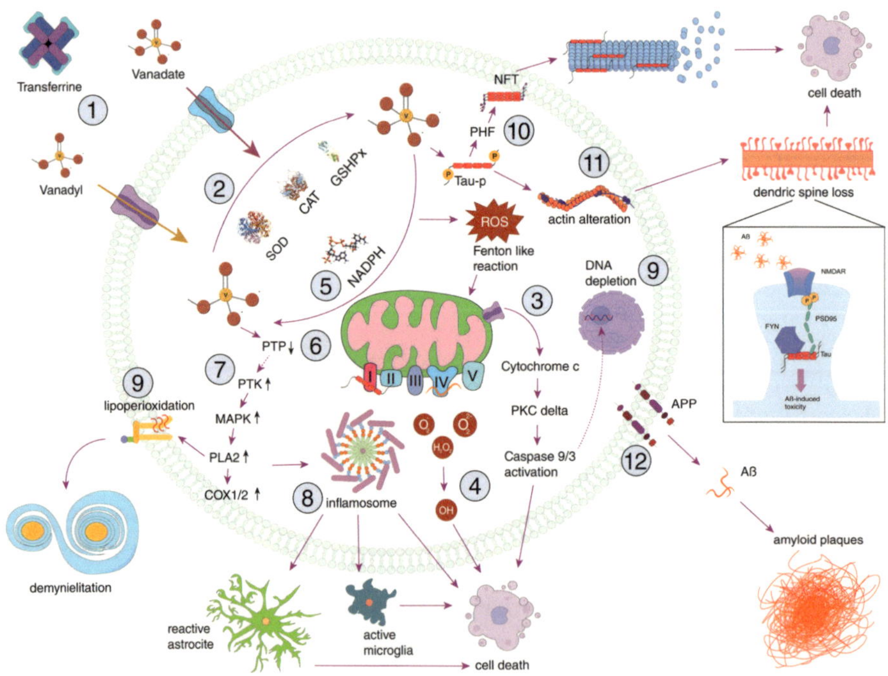


**Figure 15.**
*Possible signaling pathways of $V_2O_5$-causing AD.*

DNA and proteins alteration, lipid peroxidation, demyelination and subsequent neuronal death, demyelination and proteins damage is through lipid peroxidation [76, 78]. (10) $V^{5+}$ hyperphosphorylates tau protein (Tau-p), which leads to establishing paired helical filaments (PHF). This accumulation forms neurofibrillary tangles (NFTs), triggering microtubules alteration [54, 79]. (11) Tau-p modifies the structure of proteins in dendritic spines (actin, fyn-psd96), initiating their loss [80, 81], dendritic spines serve as crucial structures on neuronal processes, facilitating excitatory synaptic contacts. Their decline has been linked to synaptic transmission dysfunction observed in AD, highlighting the vital role they play in neural function [82]. (12) Finally, the accumulation of insoluble Aβ protein ultimately leads to the formation of Aβ plaques. We suggest that $V_2O_5$ may play a role in the progression of the disease, attributed to its capacity to induce oxidative stress and inflict brain damage through DNA modification, lipid alteration, and protein oxidation [74]. Excess ROS can be generated from mitochondrial malfunction and abnormal accumulation of transition metals such as V. The accumulation of Aβ and hyperphosphorylated tau appear to stimulate redox imbalance. The generation of oxidative stress is closely associated with the neurotoxic effects induced by tau or Aβ. Additionally, the production of ROS resulting from mitochondrial disarray is strongly linked to the progression of AD [83]. These are the products of V-based oxidative stress, suggesting a potential role for V in AD.

Some experimental genetic models of AD simulate in some way the intricate neuropathological and behavioral events observed in the disease. Our V-inhalation model replicates idiopathic AD. Nevertheless, very few animal models of sporadic AD exist, and the few that exist do not adequately simulate the gradual pathology and the main neuropathological features.

The hippocampal formation is one of the first structures to be affected by AD, inducing alterations in spatial and recent memory [84]. We utilized the T-maze test to evaluate this alteration [85]. Alterations in spatial memory correlate with cytological damage observed mainly in the hippocampus CA1 and, clearly, in the rest of the structures analyzed. Our results accord with the AD animal models outlined in the Alzforum database [86], where object recognition, spatial and working memory alterations, and cognition impairments are cited; the noteworthy discrepancy between Alzforum-reported AD models and our $V_2O_5$-inhalation model is that ours is induced or esporadic, whereas Alzforum models are knock-in or knock-out or transgenic.

In the $V_2O_5$ inhalation model, βA plaques deposition is more significant in the frontal cortex and amygdala, where we found both dβA and DcβA plaques. After the recovery period, the number of DcβA plaques increases. Dorokstar and Cols. Ref. [82] described that when the damage due to βA accumulation begins, dβA plaques are more profuse, and as the impairment intensifies, DcβA plaques (mostly produced by fibers and oligomers) accumulate in several brain structures [82]. The development and accumulation of βA generated by $V_2O_5$-exposure can act in two forms: decontrolling secretases and deregulating phosphatases. In this way, some period 4 metals such as V or Al, Fe, Cu, and Zn in their more positive oxidation states contribute to creating βA plaques; the mechanism of action consists of provoking an asymmetry in the secretases activity (increasing β and γ decreasing α), which forces APP to follow the amyloidogenic path, generating $βA_{1-42}$-increased production [87]. Vanadate produces phosphatase deregulation (diminishing protein phosphatase A2) and causes protein kinase hyperactivation (increasing MAPK) [66].

$V_2O_5$ inhalation also causes the development and accumulation of intracellular βA plaques in the pyramidal neurons of the brain nuclei examined. When βA plaques are produced, both intracellular and extracellular, in oligomeric form($βA_{40}$ prevails) and fibrillar form($βA_{42}$ predominates) [82]. Intraneuronal βA addition is similarly described in Down syndrome and at early AD stages [88]. Among the brain nuclei studied after $V_2O_5$ inhalation and after recovery, more accumulation of intracellular βA plaques is notorious in the amygdala and entorhinal cortex. In the frontal cortex of animals exposed to V, we found significant cell death of pyramidal neurons. This agrees with the descriptions that indicate that βA gathers especially in AD early stages in the neocortical regions [89].

Besides, after $V_2O_5$ inhalation, we observed pyramidal neuron intracellular NFT accumulation, mostly in the subiculum and CA1; In addition to NFT accumulation, cytoskeleton modification was observed. The neuron's cytoskeleton comprises microtubules, microfilaments, and intermediate filaments. Protein-associated microtubules initiate cross-bridges among the cytoskeleton components. Together with neuronal filaments, microtubules preserve the neuronal structure and dendritic and axonal transport, restore and adjust to pathological processes, and dendritic spine formation and functionality [90, 91]. Tau proteins are commonly found in the axon and MAPs 1 and 2 within dendrites [90]. V is a phosphate analog; in the +5 oxidation form, it adopts a tetrahedral arrangement comparable to that of phosphate [92], so it is expected that, in this oxidation form, $V^{5+}$ competes with phosphate and "vanadyl" the proteins, generating an alteration equivalent to the hyperphosphorylation of tau with the consequent damage to the cytoskeleton. Thus, chronic $V_2O_5$ exposure causes tau "hypervanadylation" (or hyperphosphorylation) that stimulates the production and accumulation of NFTs. This alteration is related to AD neural pathophysiology and the consequent spatial memory decline. It has been reported that tau pathology is AD cognitive impairment-associated and progression [5, 91].

Pyramidal neurons total number decreased by more than 40% in the frontal cortex and more than 60% in the hippocampus CA1. This corresponds with what was reported previously [93], which established that, in AD, the neocortex pyramidal neurons layers III and V and hippocampus stratum pyramidale show large loss of neurons and synapses, and the distribution of βA plaques and NFTs is more common in these cortical layers.

We also detected that $V_2O_5$ inhalation induced a noteworthy loss of dendritic spines in all brain structures examined, worsened in $V_2O_5$-exposed animals left to recover. In this respect, it has been reported that synaptic activity is altered in the brains of patients with AD due to the resulting neuronal death and loss of dendritic spines [82]. Agreeing with Dorostkar et al. [82], NFTs presence has been linked to dendritic spine loss, a connection that we also perceived. There seems to be an association between NFTs and spine number and integrity. Dendritic spines cytoskeletal instability in is associated with tau hyperphosphorylation, or, in this case, "hypervanadilation." Besides, as earlier described, $V_2O_5$ changes cytoskeletal proteins such as γ-tubulin [54], generating changes in the actin protein [94]. There appears to be a close interaction between actin and V [81], presenting great resemblance for cytoskeletal actin-binding sites. G- and F-actin interact with V [80, 95]. Additionally, V can modify actin's assembly by oxidizing its cysteines in polymerized form [80]. Moreover, the extensive neuronal loss we found in all the studied brain structures might be the consequence of the V-G-actin affinity because nerve cells involve constant actin filament polymerization due to their exceptionally dynamic cytoskeleton [96].

## 5. Conclusion

Our results demonstrate that when $V_2O_5$ is chronically inhaled, it provokes synaptic alterations, exhibited by the notable dendritic spines loss and by the presence of Alzheimer-type NFTs, mainly in the hippocampal formation, a circumstance contemplated to be the leading AD neuropathological mark, associated with the tau protein hyperphosphorylation and cytoskeletal alterations. We also found a significant presence of βA plaques and extensive neuronal damage and death in all the analyzed structures and found no recovery once the animals stopped inhaling the V compound, indicating that once the neurodegenerative process is established and homeostasis is broken, the damage continues and worsens.

Our AD model is related to Braak stage IV [97], where cognitive alterations are derived from the neuronal damage in the hippocampal formation and neocortex. This model represents an ideal experimental platform for studying sporadic Alzheimer's disease in its intermediate phase. It provides a unique opportunity to assess the efficacy of experimental drugs, given the less extensive degeneration compared to advanced stages. Furthermore, an idiopathic Alzheimer's disease model holds significant promise for evaluating and developing effective therapies. Moreover, the exacerbation of symptoms 6 months post-cessation of $V_2O_5$ inhalation underscores the significance of our model in dissecting the disease's progression. It presents a compelling opportunity to explore advanced treatments that can halt cognitive decline by impeding the formation of βA plaques or NTFs, ultimately preventing neuronal death.

Our findings will play a crucial role in advancing research on the health impacts of V and in addressing the escalating atmospheric pollution of metals such as V. This pollution has surged in the atmosphere over the past few decades, posing a significant health challenge as metal pollution is closely linked to the rising incidence of neurodegenerative diseases.

## Acknowledgements

We are very grateful to Veronica Rodríguez Mata for her excellent histologic and photographic assistance.

## Author details

Maria Rosa Avila-Costa[1*], Claudia Dorado-Martínez[1], Enrique Montiel-Flores[1], José Luis Ordoñez-Librado[1], Ana Luisa Gutierréz-Valdez[1], Cesar Alfonso Garcia Caballero[1], Javier Sanchez-Betancourt[1], Leonardo Reynoso-Erazo[2], Rocío Tron-Alvarez[2], Vianey Rodríguez-Lara[3], Jesús Espinosa-Villanueva[1] and Patricia Aley-Medina[1]

1 Neuromorphology Lab, Facultad de Estudios Superiores Iztacala, UNAM, Mexico

2 Health Education Project, Facultad de Estudios Superiores Iztacala, UNAM, Mexico

3 Department of Cell Biology, Facultad de Medicina, Nacional University of Mexico (UNAM), Mexico City, Mexico

*Address all correspondence to: nigraizo@unam.mx and avila_costa@me.com

## References

[1] Rosa E, Mahendram S, Ke YD, Ittner LM, Ginsberg SD, Fahnestock M. Tau downregulates BDNF expression in animal and cellular models of Alzheimer's disease. Neurobiology of Aging. 2016;**48**:135-142. DOI: 10.1016/j.neurobiolaging.2016.08.020

[2] Long JM, Holtzman DM. Alzheimer disease: An update on pathobiology and treatment strategies. Cell. 2019;**179**(2):312-339. DOI: 10.1016/j.cell.2019.09.001

[3] Castellani RJ, Zhu X, Lee H-G, Smith MA, Perry G. Molecular pathogenesis of Alzheimer's disease: Reductionist versus expansionist approaches. International Journal of Molecular Sciences. 2009;**10**(3):1386-1406. DOI: 10.3390/ijms10031386

[4] Braak F, Braak H, Mandelkow E-M. A sequence of cytoskeleton changes related to the formation of neurofibrillary tangles and neuropil threads. Acta Neuropathologica. 1994;**87**(6):554-567. DOI: 10.1007/BF00293315

[5] Iqbal K, Grundke-Iqbal I. Alzheimer neurofibrillary degeneration: Significance, etiopathogenesis, therapeutics and prevention. Journal of Cellular and Molecular Medicine. 2008;**12**(1):38-55. DOI: 10.1111/j.1582-4934.2008.00225.x

[6] Ittner LM, Götz J. Amyloid-β and tau — A toxic pas de deux in Alzheimer's disease. Nature Reviews. Neuroscience. 2011;**12**(2):67-72. DOI: 10.1038/nrn2967

[7] Hardy JA, Higgins GA. Alzheimer's disease: The amyloid cascade hypothesis. Science. 1992;**256**(5054): 184-185. DOI: 10.2174/1570159X15666170116143743

[8] Braak H, Braak E. Neuropathological stageing of Alzheimer-related changes. Acta Neuropathologica. 1991;**82**(4):239-259. DOI: 10.1007/BF00308809

[9] Setti SE, Raymick J, Hanig J, Sarkar S. *In vivo* demonstration of Congo red labeled amyloid plaques *via* perfusion in the Alzheimer disease rat model. Journal of Neuroscience Methods. 2021;**353**:109082. DOI: 10.1016/j.jneumeth.2021.109082

[10] Nriagu JO. A silent epidemic of environmental metal poisoning? Environmental Pollution. 1988;**50**(1-2):139-161. DOI: 10.1016/0269-7491(88)90189-3

[11] Sasaguri H, Hashimoto S, Watamura N, Sato K, Takamura R, Nagata K, et al. Recent advances in the modeling of Alzheimer's disease. Frontiers in Neuroscience. 2022;**16**:807473. DOI: 10.3389/fnins.2022.807473

[12] Hashimoto S, Matsuba Y, Kamano N, Mihira N, Sahara N, Takano J, et al. Tau binding protein CAPON induces tau aggregation and neurodegeneration. Nature Communications. 2019;**10**(1):2394. DOI: 10.1038/s41467-019-10278-x

[13] Saito T, Mihira N, Matsuba Y, Sasaguri H, Hashimoto S, Narasimhan S, et al. Humanization of the entire murine Mapt gene provides a murine model of pathological human tau propagation. The Journal of Biological Chemistry. 2019;**294**(34):12754-12765. DOI: 10.1074/jbc.RA119.009487

[14] Dujardin S, Colin M, Buee L. Animal models of tauopathies and their implications for research/translation

into the clinic. Neuropathology and Applied Neurobiology. 2014;**41**(1):59-80. DOI: 10.1111/nan.12200

[15] Drummond E, Wisniewski T. Alzheimer's disease: Experimental models and reality. Acta Neuropathologica. 2017;**133**(2):155-175. DOI: 10.1007/s00401-016-1662-x

[16] Wisniewski T, Goñi F. Immunotherapeutic approaches for Alzheimer's disease. Neuron. 2015;**85**(6):1162-1176. DOI: 10.1016/j.neuron.2014.12.064

[17] LaFerla FM, Green KN. Animal models of Alzheimer disease. Cold Spring Harbor Perspectives in Medicine. 2012;**2**(11):a006320. DOI: 10.1101/cshperspect.a006320

[18] Neff EP. Animal models of Alzheimer's disease embrace diversity. Laboratory Animals. 2019;**48**(9):255-259. DOI: 10.1038/s41684-019-0377-8

[19] Selkoe DJ. Alzheimer's disease is a synaptic failure. Science. 2002;**298**(5594):789-791. DOI: 10.1126/science.1074069

[20] Costa LG, Cole TB, Dao K, Chang Y-C, Coburn J, Garrick JM. Effects of air pollution on the nervous system and its possible role in neurodevelopmental and neurodegenerative disorders. Pharmacology & Therapeutics. 2020;**210**:107523. DOI: 10.1016/j.pharmthera.2020

[21] Hirtz D, Thurman DJ, Gwinn-Hardy K, Mohamed M, Chaudhuri AR, Zalutsky R. How common are the "common" neurologic disorders? Neurology. 2007;**68**(5):326-337. DOI: 10.1212/01.wnl.0000252807.38124.a3

[22] Olaolorun FA, Olopade FE, Usende IL, Lijoka AD, Ladagu AD, Olopade JO. Neurotoxicity of vanadium. In: Advances in Neurotoxicology. Vol. 5. Academic Press; 2021. pp. 299-327. DOI: 10.1016/bs.ant.2021.01.002

[23] Aschner M, Syversen T, Souza DO, Rocha JB, Farina M. Involvement of glutamate and reactive oxygen species in methylmercury neurotoxicity. Brazilian Journal of Medical and Biological Research. 2007;**40**(3):285-291. DOI: 10.1590/s0100-879x2007000300001

[24] Farina M, Avila DS, Rocha JBT, Aschner M. Metals, oxidative stress and neurodegeneration: A focus on iron, manganese and mercury. Neurochemistry International. 2013;**62**(5):575-594. DOI: 10.1016/j.neuint.2012.12.006

[25] Folarin OR, Snyder AM, Peters DG, Olopade F, Connor JR, Olopade JO. Brain metal distribution and neuro-inflammatory profiles after chronic vanadium administration and withdrawal in mice. Frontiers in Neuroanatomy. 2017;**11**:58. DOI: 10.3389/fnana.2017.00058

[26] Barceloux DG. Vanadium. Journal of Toxicology. Clinical Toxicology. 1999;**37**(2):265-278. DOI: 10.1081/clt-100102425

[27] Fatola OI, Olaolorun FA, Olopade FE, Olopade JO. Trends in vanadium neurotoxicity. Brain Research Bulletin. 2019;**145**:75-80. DOI: 10.1016/j.brainresbull.2018.03.010

[28] Hanus-Fajerska E, Wiszniewska A, Kamińska I. A dual role of vanadium in environmental systems—Beneficial and detrimental effects on terrestrial plants and humans. Plants (Basel). 2021;**10**(6):1110. DOI: 10.3390/plants10061110

[29] Panichev N, Mandiwana K, Moema D, Molatlhegi R, Ngobeni P. Distribution of vanadium(V) species between soil and plants in the vicinity of vanadium mine. Journal of Hazardous Materials. 2006;**137**(2):649-653. DOI: 10.1016/j.jhazmat.2006.03.006

[30] Pyrzyńska K, Wierzbicki T. Determination of vanadium species in environmental samples. Talanta. 2004;**64**(4):823-829. DOI: 10.1016/bs.ant.2021.01.002

[31] Teng Y, Yang J, Sun Z, Wang J, Zuo R, Zheng J. Environmental vanadium distribution, mobility and bioaccumulation in different land-use districts in Panzhihua region, SW China. Environmental Monitoring and Assessment. 2011;**176**(1-4):605-620. DOI: 10.1007/s10661-010-1607-0

[32] Rehder D. The role of vanadium in biology. Metallomics. 2015;**7**(5):730-742. DOI: 10.1039/c4mt00304g

[33] Duce RA, Hoffman GL. Atmospheric vanadium transport to the ocean. Atmospheric Environment. 1976;**10**(11):989-996. DOI: 10.1016/0004-6981(76)90207-9

[34] Imtiaz M, Rizwan MS, Xiong S, Li H, Ashraf M, Shahzad SM, et al. Vanadium, recent advancements and research prospects: A review. Environment International. 2015;**80**:79-88. DOI: 10.1016/j.envint.2015.03.018

[35] Xiao X-y, Yang M, Guo Z-h, Jiang Z-c, Liu Y-n, Cao X. Soil vanadium pollution and microbial response characteristics from stone coal smelting district. Transactions of the Nonferrous Metals Society of China. 2015;**25**(4):1271-1278. DOI: 10.1016/S1003-6326(15)63727-X

[36] Aragon AM, Altamirano-Lozano M. Sperm and testicular modifications induced by subchronic treatments with vanadium (IV) in CD-1 mice. Reproductive Toxicology. 2001;**15**(2):145-151. DOI: 10.1016/s0890-6238(01)00117-4

[37] Nadal M, Schuhmacher M, Domingo JL. Metal pollution of soils and vegetation in an area with petrochemical industry. Science of the Total Environment. 2004;**321**(1-3):59-69. DOI: 10.1016/j.scitotenv.2003.08.029

[38] Fortoul MR-L, Rojas-Lemus M, Rodriguez-Lara V, Gonzalez-Villalva A, Ustarroz-Cano M, Cano-Gutierrez G, et al. Overview of environmental and occupational vanadium exposure and associated health outcomes. Journal of Immunotoxicology. 2014;**11**(1):13-18. DOI: 10.3109/1547691X.2013.789940

[39] Fortoul TI, Quan-Torres A, Sánchez I, López IE, Bizarro P, Mendoza ML, et al. Vanadium in ambient air: Concentrations in lung tissue from autopsies of Mexico City residents in the 1960s and 1990s. Archives of Environmental Health. 2002;**57**(5):446-449. DOI: 10.1080/00039890209601436

[40] Li Y, Zhang B, Liu Z, Wang S, Yao J, Borthwick AGL. Vanadium contamination and associated health risk of farmland soil near smelters throughout China. Environmental Pollution. 2020;**263**(Pt A):114540. DOI: 10.1016/j.envpol.2020.114540

[41] Evangelou AM. Vanadium in cancer treatment. Critical Reviews in Oncology/Hematology. 2002;**42**(3):249-265. DOI: 10.1016/s1040-8428(01)00221-9

[42] Domingo JL. Vanadium and tungsten derivatives as antidiabetic agents: A review of their toxic effects. Biological Trace Element Research.

2002;**88**(2):97-112. DOI: 10.1385/BTER:88:2:097

[43] Paternain JL, Domingo JL, Gomez M, Ortega A, Corbella J. Developmental toxicity of vanadium in mice after oral administration. Journal of Applied Toxicology. 1990;**10**(3):181-186. DOI: 10.1002/jat.2550100307

[44] Nechay BR. Mechanisms of action of vanadium. Annual Review of Pharmacology and Toxicology. 1984;**24**(1):501-524. DOI: 10.1016/bs.ant.2021.01.002

[45] Duffus JH. Carcinogenicity classification of vanadium pentoxide and inorganic vanadium compounds, the NTP study of carcinogenicity of inhaled vanadium pentoxide, and vanadium chemistry. Regulatory Toxicology and Pharmacology. 2007;**47**(1):110-114. DOI: 10.1016/j.yrtph.2006.08.006

[46] Cseh LIL, Keith S, Taylor J. Toxicological Profile for Vanadium. Atlanta (GA): The Agency for Toxic Substances and Disease Registry; 2012. Available from: https://www.ncbi.nlm.nih.gov/books/NBK592343/

[47] Gandara RM, Soares SS, Martins H, Gutierrez-Merino C, Aureliano M. Vanadate oligomers: in vivo effects in hepatic vanadium accumulation and stress markers. Journal of Inorganic Biochemistry. 2005;**99**(5):1238-1244. DOI: 10.1016/j.jinorgbio.2005.02.023

[48] Rhoads K, Sanders CL. Lung clearance, translocation, and acute toxicity of arsenic, beryllium, cadmium, cobalt, lead, selenium, vanadium, and ytterbium oxides following deposition in rat lung. Environmental Research. 1985;**36**(2):359-378. DOI: 10.1016/0013-9351(85)90031-3

[49] Calderon-Garciduenas L, Maronpot RR, Torres-Jardon R, Henriquez-Roldan C, Schoonhoven R, Acuna-Ayala H, et al. DNA damage in nasal and brain tissues of canines exposed to air pollutants is associated with evidence of chronic brain inflammation and neurodegeneration. Toxicologic Pathology. 2003;**31**(5):524-538. DOI: 10.1080/01926230390226645

[50] Morinville A, Maysinger D, Shaver A. From Vanadis to Atropos: Vanadium compounds as pharmacological tools in cell death signalling. Trends in Pharmacological Sciences. 1998;**19**(11):452-460. DOI: 10.1016/S0165-6147(98)01257-7

[51] Mukherjee B, Patra B, Mahapatra S, Banerjee P, Tiwari A, Chatterjee M. Vanadium--an element of atypical biological significance. Toxicology Letters. 2004;**150**(2):135-143. DOI: 10.1016/j.toxlet.2004.01.009

[52] Capella LS, Gefé MR, Silva EF, Affonso-Mitidieri O, Lopes AG, Rumjanek VM, et al. Mechanisms of vanadate-induced cellular toxicity: Role of cellular glutathione and NADPH. Archives of Biochemistry and Biophysics. 2002;**406**(1):65-72. DOI: 10.1016/s0003-9861(02)00408-3

[53] Capella MA, Capella LS, Valente RC, Gefe M, Lopes AG. Vanadate-induced cell death is dissociated from $H_2O_2$ generation. Cell Biology and Toxicology. 2007;**23**(6):413-420. DOI: 10.1007/s10565-007-9003-4

[54] Mussali-Galante P, Rodríguez-Lara V, Hernández-Tellez B, Avila-Costa MR, Colín-Barenque L, et al. Inhaled vanadium pentoxide decrease gamma-tubulin of mouse testes at different exposure times. Toxicology and Industrial Health. 2005;**21**(9):215-222. DOI: 10.1191/0748233705th232oa

[55] Ramírez P, Eastmond DA, Laclette JP, Ostrosky-Wegman P. Disruption of microtubule assembly and spindle formation as a mechanism for the induction of aneuploid cells by sodium arsenite and vanadium pentoxide. Mutation Research, Reviews in Mutation Research. 1997;**386**(3):291-298. DOI: 10.1016/s1383-5742(97)00018-5

[56] Montiel-Flores E, Mejía-García OA, Ordoñez-Librado JL, Gutierrez-Valdez AL, Espinosa-Villanueva J, Dorado-Martínez C, et al. Alzheimer-like cell death after vanadium pentoxide inhalation. Heliyon. 2021;**7**(8):e07856. DOI: 10.1016/j.heliyon.2021.e07856

[57] Dorado-Martínez C, Montiel-Flores E, Ordoñez-Librado JL, Gutierrez-Valdez AL, Garcia-Caballero CA, Sanchez-Betancourt J, et al. Histological and memory alterations in an innovative Alzheimer's disease animal model by vanadium pentoxide inhalation. Journal of Alzheimer's Disease. 2024;**99**(1):121-143. DOI: 10.3233/JAD-230818

[58] Ghasemi A, Jeddi S, Kashfi K. The laboratory rat: Age and body weight matter. EXCLI Journal. 2021;**20**:1431-1445. DOI: 10.17179/excli2021-4072

[59] Deacon RM, Rawlins JN. T-maze alternation in the rodent. Nature Protocols. 2006;**1**(1):7-12. DOI: 10.1038/nprot.2006.2

[60] Schwab C, Steele JC, McGeer PL. Pyramidal neuron loss is matched by ghost tangle increase in Guam parkinsonism-dementia hippocampus. Acta Neuropathologica. 1998;**96**(4):409-416. DOI: 10.1007/s004010050912

[61] Navarro A, Valle DE, Mártínez E, Ordóñez C, Pérez C, Tolivia J. Highly selective and fast diagnosis of Alzheimer's disease Hallmark lesions using Congo red in isopropyl alcoholic solution. Journal of Alzheimer's Disease. 2013;**35**(3):589-597. DOI: 10.3233/JAD-122386

[62] The VF, Method G. A tool for comparative structural analyses. In: Nauta WJH, Ebbesson SOE, editors. Contemporary Research Methods in Neuroanatomy. Berlin, Heidelberg: Springer Berlin Heidelberg; 1970. pp. 12-31

[63] Avila-Costa MR, Colin-Barenque L, Fortoul TI, Machado-Salas P, Espinosa-Villanueva J, Rugerio-Vargas C, et al. Memory deterioration in an oxidative stress model and its correlation with cytological changes on rat hippocampus CA1. Neuroscience Letters. 1999;**270**(2):107-109. DOI: 10.1016/s0304-3940(99)00458-9

[64] Li H, Zhou D, Zhang Q, Feng C, Zheng W, He K, et al. Vanadium exposure-induced neurobehavioral alterations among Chinese workers. Neurotoxicology. 2013;**36**:49-54. DOI: 10.1016/j.neuro.2013.02.008

[65] Avila-Costa MR, Colín-Barenque L, Zepeda-Rodríguez A, Antuna SB, Saldivar OL, Espejel-Maya G, et al. Ependymal epithelium disruption after vanadium pentoxide inhalation. A mice experimental model. Neuroscience Letters. 2005;**381**(1-2):21-25. DOI: 10.1016/j.neulet.2005.01.072

[66] Ngwa HA, Ay M, Jin H, Anantharam V, Kanthasamy A, Kanthasamy AG. Neurotoxicity of vanadium. Advances in Neurobiology. 2017;**18**:287-301. DOI: 10.1007/978-3-319-60189-2_14

[67] Breit GN, Wanty RB. Vanadium accumulation in carbonaceous rocks: A review of geochemical controls during deposition and diagenesis.

Chemical Geology. 1991;**91**(2):83-97. DOI: 10.1016/0009-2541(91)90083-4

[68] Olopade JO, Toxicology JRCCti. Vanadium and neurotoxicity: A review. Current Topics in Toxicology. 2011;7:33-39

[69] Jaiswal MR, Kale PP. Mini review–vanadium-induced neurotoxicity and possible targets. Neurological Sciences. 2020;**41**(4):763-768. DOI: 10.1007/s10072-019-04188-5

[70] Garcia GB, Biancardi ME, Quiroga AD. Vanadium (V)-induced neurotoxicity in the rat central nervous system: A histo-immunohistochemical study. Drug and Chemical Toxicology. 2005;**28**(3):329-344. DOI: 10.1081/DCT-200064496

[71] Avila-Costa MR, Fortoul TI, Niño-Cabrera G, Colín-Barenque L, Bizarro-Nevares P, Gutiérrez-Valdez AL, et al. Hippocampal cell alterations induced by the inhalation of vanadium pentoxide ($V_2O_5$) promote memory deterioration. Neurotoxicology. 2006;**27**(6):1007-1012. DOI: 10.1016/j.neuro.2006.04.001

[72] Liochev SI, Fridovich I. Vanadate-stimulated oxidation of NAD(P)H in the presence of biological membranes and other sources of $O_2$. Archives of Biochemistry and Biophysics. 1990;**279**(1):1-7. DOI: 10.1016/0003-9861(90)90454-7

[73] Barth A, Schaffer AW, Konnaris C, Blauensteiner R, Winker R, Osterode W, et al. Neurobehavioral effects of vanadium. Journal of Toxicology and Environmental Health, Part A. 2002;**65**(9):677-683. DOI: 10.1080/15287390252900377

[74] Aureliano M, Sousa-Coelho ALD, Dolan CC, Roess DA, Crans DC. Biological consequences of vanadium effects on formation of reactive oxygen species and lipid peroxidation. International Journal of Molecular Sciences. 2023;**24**(6):5382. DOI: 10.3390/ijms24065382

[75] Goc A. Biological activity of vanadium compounds. Open Life Sciences. 2006;**1**(3):314-332. DOI: 10.2478/s11535-006-0029-z

[76] Irving E, Stoker AW. Vanadium compounds as PTP inhibitors. Molecules. 2017;**22**(12):2269. DOI: 10.3390/molecules22122269

[77] Tsave O, Petanidis S, Kioseoglou E, Yavropoulou MP, Yovos JG, Anestakis D, et al. Role of vanadium in cellular and molecular immunology: Association with immune-related inflammation and pharmacotoxicology mechanisms. Oxidative Medicine and Cellular Longevity. 2016;**2016**:4013639. DOI: 10.1155/2016/4013639

[78] Korbecki J, Baranowska-Bosiacka I, Gutowska I, Chlubek D. Vanadium compounds as pro-inflammatory agents: Effects on cyclooxygenases. International Journal of Molecular Sciences. 2015;**16**(6):12648-12668. DOI: 10.3390/ijms160612648

[79] Alonso A, Zaidi T, Novak M, Grundke-Iqbal I, Iqbal K. Hyper-phosphorylation induces self-assembly of tau into tangles of paired helical filaments/straight filaments. Proceedings of the National Academy of Sciences of the United States of America. 2001;**98**(12):6923-6928. DOI: 10.1073/pnas.121119298

[80] Ramos S, Manuel M, Tiago T, Duarte R, Martins J, Gutiérrez-Merino C, et al. Decavanadate interactions with actin: Inhibition of G-actin polymerization and stabilization of decameric vanadate. Journal of Inorganic

Biochemistry. 2006;**100**(11):1734-1743. DOI: 10.1016/j.jinorgbio.2006.06.007

[81] Ramos S, Moura JJG, Aureliano M. Recent advances into vanadyl, vanadate and decavanadate interactions with actin. Metallomics. 2011;**4**(1):16-22. DOI: 10.1039/c1mt00124h

[82] Dorostkar MM, Zou C, Blazquez-Llorca L, Herms J. Analyzing dendritic spine pathology in Alzheimer's disease: Problems and opportunities. Acta Neuropathologica. 2015;**130**(1):1-19. DOI: 10.1007/s00401-015-1449-5

[83] Zhao Y, Zhao B. Oxidative stress and the pathogenesis of Alzheimer's disease. Oxidative Medicine and Cellular Longevity. 2013;**2013**:316523. DOI: 10.1155/2013/316523

[84] Ameen-Ali KE, Wharton SB, Simpson JE, Heath PR, Sharp P, Berwick J. Review: Neuropathology and behavioural features of transgenic murine models of Alzheimer's disease. Neuropathology and Applied Neurobiology. 2017;**43**(7):553-570. DOI: 10.1111/nan.12440

[85] Wenk GL. Assessment of spatial memory using the T maze. Current Protocols in Neuroscience. 1998;**4**(1):8.5B.1-8.5A.7. DOI: 10.1002/0471142301.ns0805bs04

[86] Kinoshita J, Clark T. Alzforum. Methods in Molecular Biology. 2007;**401**:365-381. DOI: 10.1007/978-1-59745-520-6_19

[87] Jellinger KA. The relevance of metals in the pathophysiology of neurodegeneration, pathological considerations. International Review of Neurobiology. 2013;**110**:1-47. DOI: 10.1016/B978-0-12-410502-7.00002-8

[88] Gouras GK, Tsai J, Naslund J, Vincent B, Edgar M, Checler F, et al. Intraneuronal Aβ42 accumulation in human brain. The American Journal of Pathology. 2000;**156**(1):15-20. DOI: 10.1016/s0002-9440(10)64700-1

[89] Thal DR, Griffin WST, Vos RAI, Ghebremedhin E. Cerebral amyloid angiopathy and its relationship to Alzheimer's disease. Acta Neuropathologica. 2008;**115**(6):599-609. DOI: 10.1007/s00401-008-0366-2

[90] Avila J, Jiménez JS, Sayas CL, Bolós M, Zabala JC, Rivas G, et al. Tau structures. Frontiers in Aging Neuroscience. 2016;**8**:262. DOI: 10.3389/fnagi.2016.00262

[91] Šimić G, Kostović I, Winblad B, Bogdanović N. Volume and number of neurons of the human hippocampal formation in normal aging and Alzheimer's disease. The Journal of Comparative Neurology. 1997;**379**(4):482-494. DOI: 10.1002/(sici)1096-9861(19970324)379:4<482::aid-cne2>3.0.co;2-z

[92] Huyer G, Liu S, Kelly J, Moffat J, Payette P, Kennedy B, et al. Mechanism of inhibition of protein-tyrosine phosphatases by vanadate and pervanadate. The Journal of Biological Chemistry. 1997;**272**(2):843-851. DOI: 10.1074/jbc.272.2.843

[93] Akram A, Christoffel D, Rocher AB, Bouras C, Kövari E, Perl DP, et al. Stereologic estimates of total spinophilin-immunoreactive spine number in area 9 and the CA1 field: Relationship with the progression of Alzheimer's disease. Neurobiology of Aging. 2008;**29**(9):1296-1307. DOI: 10.1016/j.neurobiolaging.2007.03.007

[94] Rodríguez-Lara V, Morales-Rivero A, Rivera-Cambas AM, Fortoul TI.

Vanadium inhalation induces actin changes in mice testicular cells. Toxicology and Industrial Health. 2016;**32**(2):367-374. DOI: 10.1177/0748233713501364

[95] Combeau C, Carlier MF. Probing the mechanism of ATP hydrolysis on F-actin using vanadate and the structural analogs of phosphate BeF-3 and A1F-4. The Journal of Biological Chemistry. 1988;**263**(33):17429-17436

[96] Jhang KA, Park J-S, Kim H-S, Chong YH. Resveratrol ameliorates tau hyperphosphorylation at Ser396 site and oxidative damage in rat hippocampal slices exposed to vanadate: Implication of ERK1/2 and GSK-3β signaling cascades. Journal of Agricultural and Food Chemistry. 2017;**65**(44):9626-9634. DOI: 10.1021/acs.jafc.7b03252

[97] Braak H, Alafuzoff I, Arzberger T, Kretzschmar H, Tredici KD. Staging of Alzheimer disease-associated neurofibrillary pathology using paraffin sections and immunocytochemistry. Acta Neuropathologica. 2006;**112**(4):389-404. DOI: 10.1007/s00401-006-0127-z

Chapter 8

# Preclinical Models for Gastrointestinal Disease and Associated Conditions

*Fabiano Marcelo Fabris, Beatriz Gomes Pinto, Geraldo de França Júnior, Radamés Bezerra Melo, Clarissa Perdigão Melo Ferraz and Orleâncio Gomes Ripardo de Azevedo*

## Abstract

Animal models have a pivotal role in research advancing and provide cutting-edge knowledge to understand gastrointestinal diseases, offering several insights into the pathogenesis, progression, and potential treatments of conditions such as inflammatory bowel disease (IBD), enteric infection-malnutrition, obesity, gastritis, gastric and colorectal cancer, and gastrointestinal infection induced by *Cryptosporidium parvum, Clostridioides difficile,* and *Helicobacter pylori*. Rodent models are extensively used due to their genetic backgrounds like humans, ease of genetic modification, well-described immune systems, and handling. On the other hand, in IBD studies, chemically induced colitis, transgenic mice, and spontaneous disease models can replicate various aspects of human gut diseases, facilitating the study of immune responses, microbiota interactions, and genetic factors. Murine models of C57BL/6 J are extensively used for intestinal infection addressing targets and the exact molecular mechanism involved in the pathogenesis. Also, cancer and obesity models have arisen as tools to understand the cross talk between the molecule signalization that drives the metabolic modifications.

**Keywords:** animal models, inflammatory bowel disease, gut diseases, gastric diseases, intestinal diseases, rodent models, cancer

## 1. Introduction

Gastrointestinal (GI) disorders comprise a diverse group of diseases that directly affect the digestive tract. GI diseases have an annual cost calculated at about 135.9 billion dollars. The low- and middle-income countries have an even greater impact since their health system are not well-financed leading to a structural weakness and high ratios of morbidity and mortality.

Among the GI disorders, there are several pathological conditions that this chapter will shed light on. Many of them have a great impact on the health state of the population. Thus, experimental and clinical investigations could provide targets addressing diagnostics and treatment of the diseases.

Animal models are one of the most useful tools to identify the pathophysiological mechanisms involved in onset of the gastrointestinal disorders, for instance, associated pathogens, immune cell mediation, molecular recognition mechanisms, and developing therapeutic targets.

This chapter aims to present animal models that have been used as a tool to study gastrointestinal disorders and other associated conditions. Basic and preclinical research can produce knowledge that could serve as rationales to develop clinical trials, generating and providing clinical data to create and standardize the guidelines to treat and eventually cure those conditions.

One of the main pathological disorders that affects the gastrointestinal tract is the inflammatory bowel disease (IBD). The animal models used in this field focus on clarifying the immunoinflammatory mechanisms involved in the onset of the disease, for instance, the triggers activated, even though these models have their limitations they usually provide insights into how the disease progresses and in which steps treatment could be developed [1].

Furthermore, the use of animal models is very common, and the models are attributable to investigations about the development of intestinal dysfunctions and enteropathies associated with malnutrition. Malnourished rodents are used to evaluate the development of immunity in undernutrition, as well as the deficient activation of inflammatory response in these conditions. In this field, protection mechanisms for enteropathies caused by malnutrition are also targeted, through the study of peptides, cofactors, and other molecules that have this protective function, such as dipeptide alanyl-glutamine, vitamin A, zinc, and glutamine [2–7].

Furthermore, through animal models focusing on the gastrointestinal tract, it is possible to identify correlations between the vicious cycle of malnutrition, environmental enteropathy, and genetic aspects. In this aspect, studies using genetically modified mice models under malnutrition and enteric infection are discussed and could provide details about elucidating solutions.

Moreover, the use of animal models to study conditions such as obesity could be very helpful since experimental research can find mechanisms and molecular targets that may serve as targets, such as hormones, peptides, and receptors, to develop therapeutic tools to treat these conditions.

Gastric infection caused by *Helicobacter pylori* and other gastric conditions such as gastritis, peptic ulcer, and gastric and intestinal cancer also have critical relevance in the study of gastrointestinal diseases, and it is very critical to understand the mechanisms involved in proposing targets for treatments.

This chapter will focus on GI as a target system where the disorders originate; on the other hand, the gastrointestinal tract could be a system that suffers the negative impact of diseases like anxiety and depression in other systems such as the central nervous system (CNS), and some of the related aspects will be discussed.

Thus, this chapter aims to reunite several studies that showed relevance and highlighted the use of animal models demonstrating how the animal models could be very elucidative in scrutinizing the molecular mechanisms of the diseases.

## 2. Animal models for inflammatory bowel disease (IBD)

Among the disorders that affect gastrointestinal system, IBD corresponds to a clinical condition with a multifactorial cause, whose definitive pathophysiological mechanisms and the treatment tools are not yet available and whose main therapeutic strategy consists of controlling inflammation and the symptoms of the disease [8]. They have demonstrated a considerable increase in incidence and prevalence, globally, in the last 20 years [9].

The two main clinical presentations of IBD are Crohn's disease (CD) and ulcerative colitis (UC); both have similar gastrointestinal symptoms, which include diarrhea, abdominal pain, weight loss, rectal bleeding, and chronic intestinal inflammation [10–12]. **Figure 1** shows the representation of ileocolitis, one of the most prevalent forms of IBD subtype that affects the ileum in several segments.

IBD is a multifactorial disease, with the pathogenesis being frequently associated with very prevalent elements as follows: unbalanced environment, genetic background susceptibility, a significant modification in intestinal microbiota, and dysfunctional colonic immune responses [13].

IBD has been identified as predominantly affecting patients under 40 years of age, of mixed ethnicity, and living in a low-income place. Even so, it is a globally distributed condition that affects a wide range of people [14].

Despite the recent significant increase in IBD in the global context, the methods for noninvasive study of these conditions are still limited and lead to a gap in the knowledge of the mechanism of disease. With a multifactorial cause, IBD has been studied aiming to elucidate its pathophysiology by identifying treatments and therapies that improve the quality of life of the affected population and aiming to achieve total remission of symptoms and cure [15].

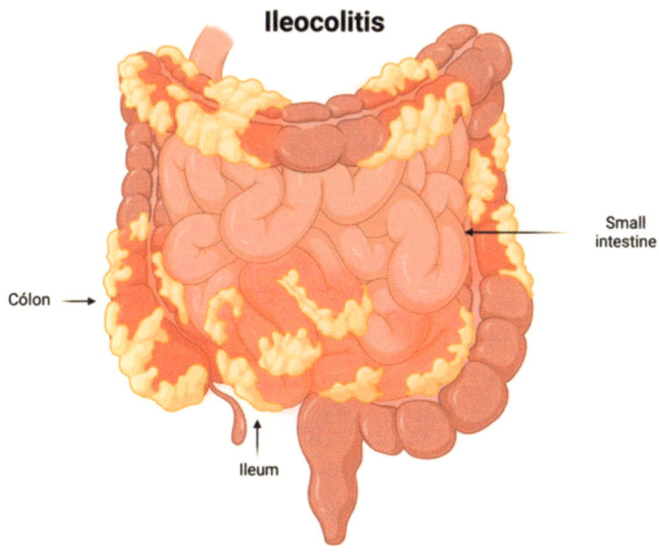


**Figure 1.**
*Schematic representation of ileocolitis, showing the macroscopy signs of inflammatory areas. This kind of IBD subtype is one of the most prevalent conditions of CD, highlighting the intestinal portions that could be affected. Created with BioRender.com.*

The main symptomatology of IBD consists of chronic intestinal inflammatory aspects, with several worsening periods [9]. CD is characterized as an intestinal inflammation that can affect any region of the gastrointestinal tract, from the oral cavity to the anus. In most cases, CD affects the small intestine and the beginning of the large intestine. On the other hand, UC is characterized as an inflammation that is more prevalent in the colon [16].

Both pathologies result from dysfunctions of the immune system, environmental and genetic factors, resulting in an exacerbated inflammatory response to intestinal pathogens and other pro-inflammatory substances in a genetically susceptible patient [9]. It is estimated that approximately 200 *loci* of the genome are related to the pathogenesis and maintenance of IBD [17].

The symptoms are similar and very frequent and the treatment is also similar and involves controlling the symptoms and the inflammatory process, as well as changing eating habits, and eventually the patients are refractory to some of these treatments and need to treat with immunobiological or even surgical procedures [16].

In some severe cases, the patient may also experience mineral salt deficiencies and hypovitaminosis, which compromises several metabolic mechanisms [6, 18].

The main factors that can trigger IBD are related to continuous exposure to stress. The emotional condition could lead to an exacerbation of inflammatory response. In the same way, pathogenic bacteria can increase the inflammatory state. In addition, infection by Gram-negative bacterium, for instance, the *Vibrio cholera* in the intestine, which is associated with high levels of pro-inflammatory biomarkers, and also foods and other toxic compounds that are orally administered can trigger this inflammatory pattern [19].

Furthermore, the importance of the availability of animal models that mimic the human gut in laboratory research involving the mechanisms related to the pathogenesis of IBD is crucial. Through studies using specific animal models, it is possible to improve the understanding of the pathogenesis of gastrointestinal disorders, as well as their specificities. Furthermore, it is also possible to identify the best treatment methods and their limitations based on physiological and pharmacological studies [20].

These models also provide details about pharmacological interactions that could address drug development and, in addition, could provide data to improve the molecular immune mechanisms in immunobiological therapy, focusing on elucidating the interplay between mucosal epithelium, intestinal microbiota, and immune interactions [21–24]. Despite other animal categories, mainly mice and rats are the most predominant models used; also other mammals and nonmammals are adopted for IBD modeling [25–29].

Other animals such as monkeys frequently exhibit spontaneous colitis, bearing a greater resemblance to human patients [30]. The modeling methods include chemical induction, genome modifications, and intestinal microbiota modulation [25, 31].

Among the many models used to study IBD, rodent models are frequently adopted since they comprise the basic characteristics needed. The rat species Sprague-Dawley (SD) and Wistar are the most used animals. The methods used frequently are chemical compounds in drinking water with acetic acid, trinitrobenzene-sulfonic acid (TNBS)-ethanol mixture, dextran sulfate sodium (DSS), and oxazolone with similar methods for Wistar [20, 25, 31–34].

On the other hand, in mice, C57BL/6 J and BALB/c strains are the most frequent mice species used, due to the feasibility of handling and managing the protocols. Chemical induction is one of the most common induction methods chosen and currently DSS and 2,4,6-tTNBS are the most widely used chemical compounds to induce IBD [32, 33, 35, 36]. In addition, studies suggested that the DSS model is closely related to human UC, while the TNBS-induced model is more related to CD based on immunological mechanisms.

Conversely, the animal model could be developed based on genetic modifications. These modifications in mice are very usual methods, and the genetical changes can include knocking out IL-10 [37, 38], SAMP1/YitFc [39], and T-lymphocyte receptors, and transgenic models for genes that express interleukin-7 [17, 24]. Beyond the genetic modification, another strategy for the gastrointestinal inflammation model it is T lymphocytes transfer. In study by Powrie and cols, the authors made a significant advance in understanding of animal's models for intestinal inflammation, transferring CD4+ T cells, from a lymphopenic mice donor, and consequently inducing a worsening of gastrointestinal inflammation on recipient, after 5–10 weeks of the treatment (10.1093/intimm/5.11.1461.); also, genetic modification in IL-23 receptor in immune cells can drove the inflammatory state and promotes colitogenic activity (10.1016/j.immuni.2010.08.010) [40].

Also, the model can address peptides Mdr1a$^{-/-}$, N-Cadherin, and NF-κB, essential modulators for knockout models that present epithelial dysfunction and are suitable for exploring intestinal epithelial cell–related onset of disease [41–44]. Furthermore, genetic modifications in mice can lead to immune dysfunction, ultimately resulting in the development of colitis.

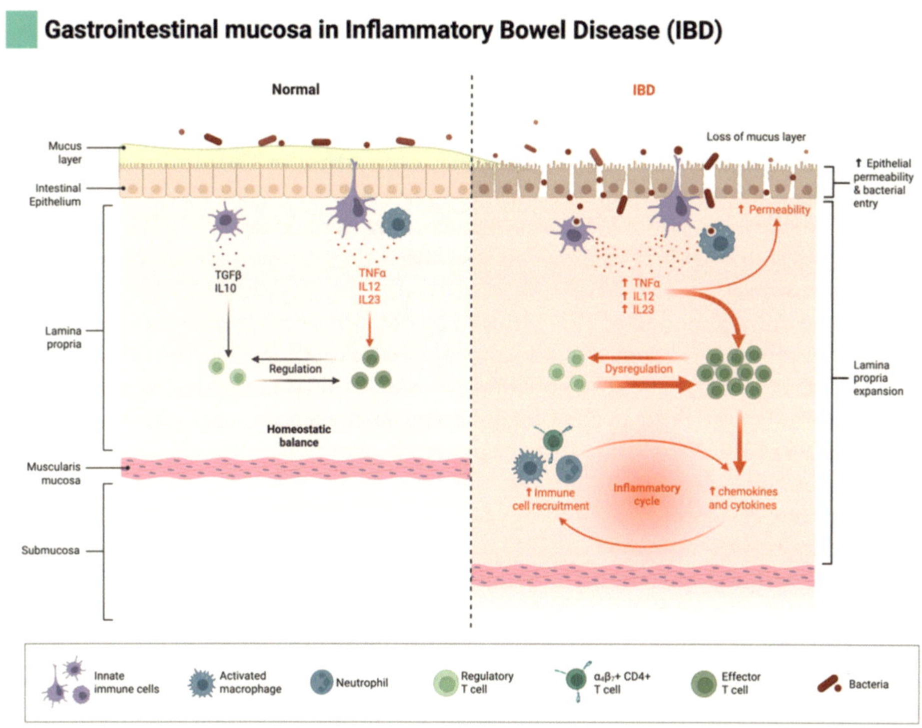


**Figure 2.**
*Schematic representation adapted [45] from enteric infection highlighting the immune mechanisms involved in IBD showing damage of the intestinal barrier, leading to increased intestinal permeability, loss of mucus layer, lamina propria expansion, elevation on the pro-inflammatory cytokines release associated to the TNF-α activity elevation, and disarrangement in the tissue architecture. Created with BioRender.com.*

It highlights the pivotal role of enteric infectious and intestinal dysbiosis in breaking the intestinal barrier and consequently increasing intestinal permeability, a condition known as "leaky gut". Addressing these situations, there are animal models pre-exposed to infections by microorganisms related to the development of IBD, such as *Trichinella spiralis, Campylobacter jejuni, Salmonella enterica,* and pathogenic strains of *Escherichia coli* [8]. **Figure 2** highlighted the leaky gut in comparison to the healthy gut demonstrating the profound differences between the gastrointestinal epithelia when the enteric infection is associated with malnutrition.

Furthermore, a model of colon inflammation in mice induced through enteric infection and hypothermia was also developed. For this method, the authors fed a newborn mouse with a formula, which could contain *Escherichia coli*, then exposed it to nitrogen gas for a few seconds, and kept it in an environment of approximately 4°C for 10 minutes. This causes vascular contraction and reduces intestinal circulation, leading to an inflammatory necrotizing tissue lesion of the colon [46].

**Table 1** summarizes the studies highlighting the main patterns associated with the experimental studies of IBD in animal models, such as induction method, which was the main disease investigated showing advantages and disadvantages of each of them.

However, these experimental models could not investigate the disease in the whole aspects related to the pathogenesis. Even so, despite the limitations of each type of induction, these models help to understand the pathophysiology mechanisms and maintenance of the disease and to test the effectiveness of new treatments and new pharmacological targets [1].

Studies in animal models induced with IBD have allowed the identification of the main epigenetic factors, linked to habits, that contribute to the development of IBD. These include maternal habits, such as diet, infectious conditions, smoking during pregnancy and breastfeeding, and personal habits, such as microbiota health, a diet rich in high fat and low fiber, vitamin deficiency, and use of antibiotics [48].

In addition to understanding the mechanisms that trigger IBD, animal models have also allowed to understand factors that can protect or aggravate the inflammation, such as the protective role of vitamins A and D in mice against IBD induced by chemical compounds [49]. From this perspective, other protective factors are related to the diet consumed daily.

Studies in animal models of IBD induced by chemical compounds have identified an improvement of IBD in mice that received specific diets, such as the fasting-mimicking diet (FMD). The study by Rangan, et al. evaluated the dietary effects of four-day-fast-mimicking cycles diet on dextran sulfate sodium (DSS)-induced murine model of IBD. The study found a beneficial effect on reducing the intestinal inflammation, increased the number of intestinal stem cells, and stimulated a protective microbiota, reversing the pathology induced by DSS [50].

Furthermore, animal models of IBD contribute to understanding the relationship between the immune system and intestinal microbiota. It was shown that inflammation is suppressed in mice with healthy microbiota, while mice without microbiota (germ-free, GF) were unable to suppress the inflammatory condition [17]. In a study by Paik et al., the authors have demonstrated that GF Smad3$^{-/-}$ mice with disrupted TGF-β signaling pathway are more susceptible to develop GI inflammation in a fecal transplant protocol from IBD mice, suggesting that the microbiome is an independent risk factor contributing to IBD onset [51]. In another study, Britton et al., when assessing the influence of intestinal microbiota from an IBD patient to a GF mouse, found the colonization with IBD microbiota induced an exacerbated colitis evidenced by an elevation on Th17 and RORγt$^{+}$ Treg cell counting [52].

| Induction method | Method used | Disease investigated | Advantages | Disadvantages |
|---|---|---|---|---|
| Chemical-induced colitis | Oxazolone colitis | Ulcerative colitis (UC) | • The model proposed a role for NKT cells during UC.<br>• Inflammation in the anatomical position that closely resembles UC. | • High mortality.<br>• In clinical trials, the therapeutics targeting IL-13 were not significant. |
| | Trinitrobenzene-sulfonic acid (TNBS) colitis | Crohn's disease (CD) | • The model highlighted the importance of IL-12 and IL-23 in successful clinical trials.<br>• Understanding the genetic determinants (e.g., NOD2).<br>• Understanding of host-gut microbial interactions in disease.<br>• Mice developed chronic transmural colitis (mimics of CD). | • Variation in responses requires increase of N of animals.<br>• High variability on TNBS effect. |
| | Dextran sulfate sodium (DSS) | UC and CD | • Th1/Th2 cytokine milieu closely resembles both UC and CD.<br>• The model revealed the importance of innate immunity, IL-17, and IL-23 in disease induction. | • Interbatch variability of DSS affects colitis phenotype.<br>• The disease is highly dependent on the microbial status.<br>• Repeated administration of DSS. |
| Spontaneous colitis | Ikk-γ (NEMO) deficiency colitis | NEMO-deficiency | • The model has enabled research into the influence of the microbiome on intestinal inflammation. | • NEMO-deficient patients are prone to other GI diseases. |
| | IL-10 deficiency colitis | Childhood IBD | | • IBD patients are not always deficient in IL-10 and may not benefit from IL-10 therapy. |
| Immune cell colitis | $CD4^+$ T-cell transfer | CD | • Revealed the importance of Treg cells in IBD.<br>• Allows early events underlying IBD to be studied.<br>• Supported a role for IL-17, IL-23, and NK cells in IBD development.<br>• Inflammation in small bowel and colon like CD. | • "Leakiness" and altered NK function in immune-deficient mice.<br>• Costly model.<br>• The use of immune-deficient mice isolates other factors of CD induction/development. |

**Table 1.**
*Studies highlighting the induction method used and the advantages and disadvantages of each one respectively, adapted from [47].*

To assess which animal and induction model should be chosen for a specific study addressing IBD features, several aspects must be observed; clinical observations like disease activity index score, pathological observation, intestinal epithelial structure, intestinal microbiota, inflammatory markers, and other features found in each model must be analyzed. Based on these patterns, it is possible to identify which animal model should be used according to the research objective.

Despite IBD having higher incidence and prevalence levels in the developed world, in low- and middle-income countries, several other gastrointestinal conditions prevail. The infectious disease is one of the most prevalent conditions that affects the population in these areas.

## 3. Animals models in gastrointestinal infectious disease/malnutrition, and obesity

There are a range of gastrointestinal disorders, that can have the knowledge improved by understanding through studies using animal models. To achieve this objective, it is essential to understand the main animal models used according to the pathology to be analyzed, as well as the mechanisms of development of gastrointestinal disorders used in a laboratory environment.

Gastrointestinal tract infections are caused by microorganisms, whose pathogen-associated molecular patterns (PAMPs) are identified by the immune system's recognition mechanisms, initiating an immune response. According to the World Health Organization, there were around 1.5 million deaths by diarrheal diseases worldwide in 2019, of which the most affected group was children under 5 years [53].

Although intestinal infections caused by viruses and parasites can occur, the main microorganisms related to intestinal infections are bacteria. The principal pathogenic bacteria related to pathogenesis of intestinal infection belong to *Enterobacteriaceae* family, as pathogenic strains of *Escherichia coli*, *Citrobacter freundii*, *Klebsiella pneumoniae, Proteus mirabilis*, *Morganella morganii*, *Salmonella enterica,* and *Shigella sonnei*, as well as Gram-positive bacteria such as *Staphylococcus aureus* and other Gram-positive pathogenic bacteria and other protozoa, for instance, *Cryptosporidium parvum* [54]. In addition, *Helicobacter pylori* is the number one infectious pathogen associated to gastric cancer worldwide and is related to gastritis and peptic ulcers.

Studies aiming at the pathogenesis and pathophysiology of intestinal infections using animal models provide *in vivo* information on the pathogenic routes that each microorganism possesses, as well as the virulence mechanisms it presents. Furthermore, it is also possible to understand correlations between gut infection and immune response, interactions between pathogen and intestinal microbiota, and experimentation of new antimicrobial drugs *in vivo* [1].

However, just as several neglected tropical diseases have low support for laboratory studies, research in intestinal infections is not supported. Furthermore, the use of alternative research methods without using animal models may be more favorable depending on the type of study developed.

Animal models induced with gastrointestinal infection acquire common conditions, such as dysbiosis of gut microbiota, which alters important mechanisms for proper intestinal function. By generating dysbiosis, intestinal infections affect organ sensitivity, permeability, and motility; stimulate inflammatory response; and alter immune function and enteric nervous function, culminating in the development of functional gastrointestinal disorders [55].

Furthermore, it is also possible to investigate strains of bacteria with high expression of virulence mechanisms, based on gene expression or suppression, and their antagonistic interaction with the rodent microbiota. The investigation of the pathophysiology of intestinal infections in murine models helps understand the pathogen virulence and toxigenic mechanisms.

In this regard, severe intestinal infections, such as those caused by *Vibrio cholerae* and *Enterobacteriaceae*, can be studied in rodents to identify the tissue targets of the pathogen and its mechanisms of virulence, adhesion, and intestinal colonization [19].

Similarly, studies of strains of nonpathogenic bacteria in mice are performed to evaluate the probiotic and antibacterial potential role of these bacteria to pathogenic microorganisms. In this perspective, studies using an *Escherichia coli* nonpathogenic strain showed that it exerted an antibacterial function in intestinal infections induced by *Pseudomonas aeruginosa*, inhibiting its adhesion and virulence mechanisms. These studies evaluate the replacement of antibiotics by other mechanisms that do not generate bacterial resistance and do not cause dysbiosis, as is evidenced in the use of pharmacological antibacterials [19, 56].

Furthermore, studies with murine animal models in intestinal infections may be useful in understanding protective factors against pathogens, such as enzymes, proteins, and cofactors. Proteins with immunoinflammatory activity, such as apolipoprotein E, have protective action against infectious diarrheal diseases. Studies have identified that this protein reduces weight loss in undernourished mice, helps reduce microorganisms in intestinal infection models, dissipates the infectious condition promptly, helps restore intestinal epithelial layers, and modulates the release of pro-inflammatory cytokines in infectious conditions [57].

By identifying enzymes, proteins, and cofactors that have protective action against infections, it is possible to discover new methods of reversing infections. Regarding this protection role, it was identified that ascorbate, the active form of ascorbic acid, plays a protective role in intestinal infections with *Shigella*. These studies serve the basis for understanding specific mechanisms that act in the body's defense against infections, in addition to identifying animal models in which possible treatments and methods of preventing infections can be studied [58].

Intestinal infection in rodents can be induced using a variety of protocols, by different pathogenic microorganisms. Among the most used protocols, contamination of the animal model with larvae of the parasite *Trichinella spiralis*, *Nippostrongylus brasiliensis,* and *Cryptosporidium parvum* is frequently adopted. Studies involving other microorganisms, such as *Campylobacter rodentium*, *C. jejuni*, *Salmonella enterica,* and *E. coli*, are also found for intestinal infection induction [1].

In studies investigating the pathogenesis and immunogenic interaction of microorganisms like *Vibiro cholerea,* the most used model. The bacterium is orally given to mice with 3–5 days old. This model can be used to investigate the virulence genes and aspects on bacterium colonization. On the other hand, the main limitation is the shortage of biological material for omics investigation (e.g., genomic and transcriptomic) [19].

Moreover, studies evaluating the antimicrobial role of probiotics against pathogenic bacteria use mice aged 6 to 8 weeks, in the weaning phase, that underwent intestinal infection induction via gavage. The engineered probiotic given to mice shows *in vivo* prophylactic and therapeutic activity against *P. aeruginos*a during gut infection in murine and *C. elegans* models, suggesting the strong role of probiotics controlling the pathogens community in GI [56].

The gastrointestinal tract is a system where the organism absorbs the molecules to synthesize the energy, through the digestion and metabolization of macromolecules such as carbohydrates, polypeptides, and lipids. Many disease conditions have the gut as the main site where they earlier develop. Also, as mentioned before, the GI tract could be affected by several conditions from nutrition to infectious diseases.

In 2022, 390 million people were in a state of undernutrition. Furthermore, half of deaths in children younger than 5 years in the world are related to malnutrition [59].

The importance of investigating the correlation between malnutrition and obesity in gastrointestinal tract disorders is evident. The use of animal models in malnutrition associated with enteropathies is crucial, since these studies produce knowledge to understand how to overcome this condition, but few studies correlate these factors. Most studies on malnutrition involving animal models seek to identify the relationship between a diet deficient in proteins and micronutrients, delayed development, and microbiota impact [5, 60].

Malnutrition could be defined as a deficiency, excess, or imbalance in the intake of energy and/or nutrients. The term covers undernutrition, which includes stunting, wasting, being underweight, and nutrient deficiency associated with adverse effects on composition, function, and clinical outcomes [61]. On the other hand, in obesity despite the high body mass index (BMI), the individual could be severely affected by a reduction in levels of critical molecules such as amino acids and peptides, and the adipose tissue could be a source of pro-inflammatory molecules release in those carriers [62].

The main complications of undernutrition include hypovitaminosis, lack of minerals, and essential amino acids. Most common hypovitaminosis conditions included hypovitaminosis A, which frequently leads to ophthalmic disorders; hypovitaminosis D, which is associated with calcium bone disorders; and hypovitaminosis of complex B leading to neurological, dermatological, and digestive disorders [18, 63, 64].

The main deficiencies of mineral salts like iron, iodine, and calcium deficiency cause anemia, goiter, and bone and neuromuscular disorders; also deficiency in essential amino acids lysine, methionine, and tryptophan can be found [65–67].

Studies in malnutrition using animal models are pivotal to understanding the communities' phenomena, and the studies can provide details addressing the targets on mechanisms involved.

Rats and mice have been used for evaluating the effect of malnutrition widely, and specific strains of these rodents are used. Rodents have a similar genome to humans, which gives them physiological and biochemical processes almost identical to human beings. In addition, they have a vast proliferative capacity, which allows researchers to monitor generations of rats and mice in their experimental studies [68].

The development of malnutrition in animal models, in general, occurs through modified diets usually with low amounts of protein. The C57BL/6, BALB/C weanling mice strains, and Wistar rat strains are the most used strains adopted in studies of malnutrition.

To achieve undernutrition, animals can be fed by adopting nutrient measures of malnutrition. Among nutrients consumed, a deficit of protein consumption associated with an increased rate of carbohydrates can be adopted, resembling the Western diet. These animals have their weight and growth monitored and compared to the weight and growth of animals nourished with a regular diet at the same age. The studies showed the malnourished mice demonstrated below-expected growth levels and an increased weight compared to mice under the regular diet [69, 70].

In addition, the animals can be genetically modified to develop more pronounced or faster malnutrition. However, this genetic model of malnutrition does not resemble the developed human malnutrition context, because it is completely conditioned by dietary deficiencies [5].

Here we showed some examples in how the use of this tool can address the phenomena and find targets that could be targeted to understand the mechanisms involved.

Contextualizing the animal models for clinical disease presentations, Oria et al. found that children with heavy-burden diarrhea, living in a precarious environment harm physical and cognitive development. In addition, the authors found a high prevalence (13.4%) of the APOE4 gene suggesting that the gene could protect the kids [71] since this gene is associated with a pro-inflammatory state widely recognizable in CNS of elderly people and frequently associated with a condition such as Alzheimer's. In a later study, the authors examined whether APOE4 affects the intestinal barrier function in a supplementation test with retinol, zinc, and glutamine. The outcomes showed significant associations between vitamin A, glutamine, and APOE4 background in the recovery of the intestinal barrier [72].

Moreover, Azevedo et al. in a study investigated whether the APOE genotype could be associated with a pro-inflammatory response to malnutrition and infection by protozoa. The study evaluated the association between a murine mice model of malnutrition and enteric infection, induced by a low-protein diet (2% of protein), in association with an enteric infection induced by oral gavage of $10^7$ unexcysted oocysts of *Cryptosporidium parvum* to the C57BL6J targeted replacement mice for APOE gene (APOE2/2, APOE3/3, and APOE 4/4), APOE $^{-/-}$, and wild-type ones.

APOE4/4-TR mice had better weight gain after infection plus malnutrition compared with APOE3/3-TR and wild-type mice. APOE4/4-TR and APOE$^{-/-}$ mice shed less; nonetheless, the latter showed villus blunting and higher ileal pro-inflammatory cytokines and iNOS mRNA levels. Also, APOE4/4-TR mice had increased ileal CAT-1, arginase-1, and TLR-9 mRNA levels compared to APOE$^{-/-}$ [57], critical for tissue recovery and immune-mediated defense.

The study found that the human APOE4 gene showed a protective role against the enteric infection caused by *Cryptosporidium parvum* infection aggravated by the malnutrition state, ensuring previous reports that found a state of protection against diarrhea in APOE4 children living in a Northeast shantytown of Brazil.

The animal studies could help to understand the mechanism associated with the impacts on nutrient deficiency diseases and metabolism addressing the development of new interventions and treatments in undernutrition cases.

In addition, the animal models could be a plentiful tool to understand malnutrition even in earlier stages of life. In an experimental study by Ueno et al., the authors investigated whether alanyl-glutamine (Ala-Gln) could provide any beneficial effect on the intestinal barrier in 10-day-old suckling C57BL/6 pups of malnourished mice, induced by an isocaloric "regional basic diet (RBD)".

Pups under the RBD-fed dams demonstrated failure to thrive, and also the intestinal specimens showed decreased villous/crypt histology outcomes, transmucosal resistance, and epithelial proliferation cells. On the other hand, the group showed increased permeability, the tight junction claudin-3 expression, and increased epithelial apoptosis, which were reverted in undernourished pups supplemented with Ala-Gln, demonstrating the relevance of key features of murine models of the human condition [4].

Additionally, Oria et al. exploring the potential interactions among environmental enteropathy, malnutrition, and neuroinflammation published an article using C57BL/6 weanling mice challenged by *Cryptosporidium parvum* (induced by inoculum with $10^7$ oocysts in 100 μl of PBS orally given) and undernutrition (induced by 2% of protein diet). The authors demonstrated an elevation of brain inflammation through the evaluation of inflammatory biomarkers such as NF-kB, IBA, and MPO, in the prefrontal cortex of protein-deficient and infected mice reinforcing the importance of the gut-brain axis and how the environmental enteropathy affects the central nervous system [73].

Also, in a study by Ribeiro et al., the authors evaluated the effects of the murine model of malnutrition induced by an RBD in physical development and the intestinal morphofunctional barrier. The study demonstrated that mice fed with RBD showed significantly reduced weight, increased ease in the villus: crypt ratio, and reductions in mRNA levels of the tight junction claudin-2 and occluding leading to commitment in the intestinal barrier functionality [74].

Moreover, in a study investigating the effects of malnutrition-induced ban on RBD on liver function, the authors induced a malnutrition state compared to a regular diet group. The study found a significant impairment in body weight compared to the control group and also higher RNA levels TNF-α, ApoA, and IL-10 for the undernourished group compared to controls, suggesting the malnutrition state could be associated with liver inflammation induced by RBD [75].

Furthermore, in a study discussing the immunoinflammatory role of ApoE4 in malnutrition and enteric infections, the authors discussed how the ApoE is involved in inflammatory modulation and how it influences the immune response at the cellular level. In addition, they discussed how the ApoE mimetic peptides could modulate the immune function in the gastrointestinal tract microenvironment.

The study also focuses on how the APOE4 is associated with improvements of enteric infection associated with malnutrition in heavy contaminated environments, discussing the inflammatory mechanisms; it also highlights the interaction of APOE4 genotypes impacts on the diversity of intestinal microbiota, which is a critical factor for the control of tissue inflammation of gut. Moreover, the association of E4 alleles in intestinal barrier dysfunction is discussed, which leads to a bacteria translocation and worsening of local and systemic inflammation [76].

The gastrointestinal tract is not just a system where the models could address the disease mechanisms suggesting targets for treatment and new therapeutics. The digestive system can be used as a tool to administer drugs and supplementations addressing many conditions such as the effects of administration of molecules through the gastrointestinal tract in the CNS.

Moreover, using animal models to study environmental enteropathy in association with malnutrition may also offer a fair analysis of the mechanisms involved in the brain-gut axis, addressing the elucidation of molecules and receptors involved, which is important for understanding some of the gastrointestinal disorders related to the nervous system. In this regard, some studies seek to understand the pathogenesis of eating disorders using animal models, such as anorexia, as well as the implications of stress and other conditions on intestinal health modulation [77, 78].

In a Brazilian study conducted by Barbosa et al., the authors tested whether administration of omega (ω)-9, ω-3, and ω-6 diluted in milk can prevent the oxidative alterations associated with behavioral and cognitive age-related disorders.

The study used 28-day-old male mice that received *ad libitum* skim milk enriched with omega oil compared to the control that received water in the same volume. The treatment lasted between 10 and 14 months, and the study outcomes showed that thiobarbituric acid reactive substances (TBARS) levels were decreased in the hippocampus and striatum in the 10-month enriched with omega oil mixture (EM) group compared with the control. A similar decrease rate was observed in the striatum of the 10-month skim milk (SM) group. Also, glutathione (GSH) levels were higher in all 14-month groups treated with omega oils, compared with 10-month groups. In the oxidative aspects, myleoperoxidase (MPO) activity was higher in the 14-month EM group compared to the control and SM groups, revealing a possible pro-inflammatory status [79].

This study mimics the oxidative stress associated with aging. The outcomes suggested that the GI tract can be a pathway that can be used to administer fatty acids that can contribute to reducing the oxidative stress by ameliorating the MPO and TBARS levels of mice supplemented with milk enriched in omega oils.

Another very critical condition with alarming levels worldwide is obesity. This condition has associations with the development of gastrointestinal disorders. This is due to the consolidated correlation between the development of obesity and the exacerbation of a pro-inflammatory state.

According to the World Health Organization, in 2022, around 1 in 8 people in the world would have met the criteria for obesity [80]. This condition is characterized as a complex chronic disease, due to the multifactorial pathogenesis of the condition.

The main causes correlated with the development of obesity are excessive consumption of high-calorie foods, frequently low levels of education, poor environment, reduced social interactions, some aspects linked with epigenetics, and a sedentary lifestyle [68, 81–85]. **Figure 3** shows the main causes associated with obesity.

Furthermore, a study of obese patient genotypes identified around 300 genetic polymorphisms associated with the development of obesity, but the determining

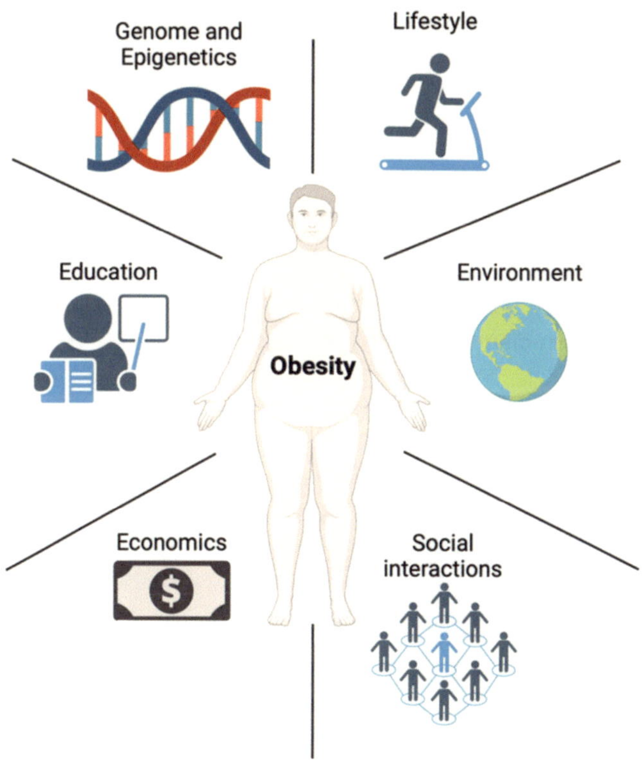


**Figure 3.**
*The main causes associated with the onset of obesity in adult individuals. All the factors are associated with obesity, leading to individual care.*

factor for the comorbidity emergence is the patient's lifestyle. Thus, patients at high risk of developing genetic obesity can inhibit this risk by adhering to a healthy lifestyle [68, 81, 83–85].

Obesity is considered a public health disorder especially due to its associations with metabolic syndrome development, such as hypertension, dyslipidemia, type 2 diabetes mellitus, chronic kidney disease, and nonalcoholic fatty liver disease, among others. Furthermore, it is known that obesity affects intestinal dysbiosis and therefore depression and anxiety, directly interfering with the gut-brain axis [83–85].

In addition, obesity favors the metabolism of reactive oxygen species in the body, increasing cell damage due to oxidative stress, which leads to the activation and release of pro-inflammatory cytokines in the organism [68, 86].

Animal models that reproduce obesity are developed, from the modification of food offered to the animal (e.g., high fat or caloric diet) and through gene alterations, with the use of modified targeted replacement genes or knockout animal models. However, few studies investigate the relationship between obesity and the development of specific inflammatory diseases of the digestive tract [86].

Therefore, the study of obesity using animal models is pivotal for expanding the pathophysiology understanding of the disease and its correlation with the described disorders.

Experimental studies can use monogenic or polygenic animal models. Monogenic models are developed from the mutation of a specific gene, or a set of specific genes, in embryonic stem cells. In this case, the animal is called "knockout" when the selected genes are deleted from their genome, while animals with genes added to their genome are called "knockin" [68, 87].

The main monogenic animal models are (1) the C57BL/6 J mouse with a mutation in the leptin gene, called *ob/ob* mouse, or "obese" mouse; (2) the C57BL/KS mouse with a spontaneous mutation in the leptin receptor gene, called *db/db* mouse, or "diabetic" rat; (3) knockout mice of MC4R gene, responsible for production of melanocortin-4 receptors; (4) NSY mice with a mutation in yellow Agouti gene, which also participates in melanocortin signaling; and (5) knockout of ApoE gene mice, leading to high levels of lipids on blood [68, 82, 88, 89]. **Figure 4** depicts the most used methods to induce obesity in mice.

Polygenic animal models undergo induced obesity originating from modified diets offered to animals. Two main types of hypercaloric diets involve the consumption

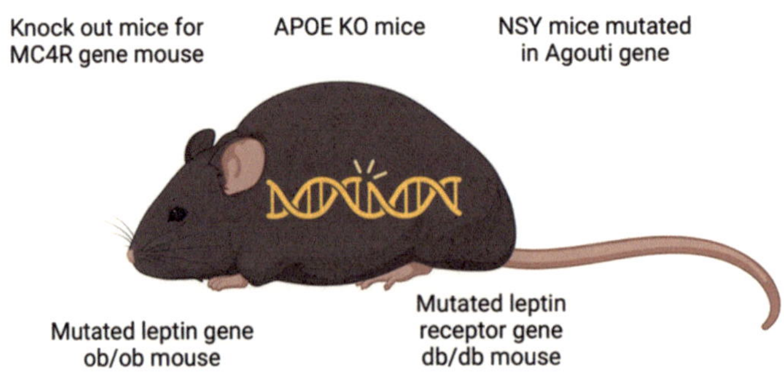


**Figure 4.**
*The figure depicts the main methods currently used to induce obesity in mice. The APOE knocking out mice develops obesity spontaneously, but frequently the authors feed these mice with a caloric and lipid-rich diet. Other types of genetic modifications can be addressed, for instance, knocking out MC4R, leptin, or leptin receptor genes. Created with BioRender.com*

of high caloric and lipid content foods or the consumption of outrageous palatable foods, stimulating hyperphagia in rodents. Animal models whose obesity is induced by diet are considered the most reliable models for human obesity studies and are usually chosen in research with preclinical, pathophysiological, and interventional analysis [68].

The main rodent models of diet-induced obesity are *Mus musculus* mice, whose strains most used in the laboratory are Kunming, C57BL/6, BALB/c, and ICR/HaJ. Of these, Kunming and ICR/HaJ strains have been shown to gain adipose tissue easily in the laboratory compared to C57BL/6 and BALB/c strains [90].

Furthermore, other diseases could be associated with the obesity animal model and can be developed together; for example, type 1 and 2 diabetes are currently used in association with methods for obesity development, and for instance, drug-induced and high-calorie diet is frequently used. In these chemical models, hyperglycemic drugs, such as streptozotocin and alloxan, are used to identify the association between obesity and the development of type 2 diabetes mellitus [91–93]. Also, gestational diabetes could be addressed by an animal model and can be studied since it is a very prevalent condition, which is associated several times with obesity [94].

Moreover, the studies could comprehend modifications in the animal habitat that can be adopted to stimulate obesity. Removing activity wheels and modifying the temperature for a chronic cold (e.g., 21°C) [95] body temperature, which stimulates metabolism and food intake by the rodents, are also highlighted.

Additionally, obesity animal models can be developed through surgical intervention, through damage to brain regions responsible for hunger and satiety [96–98]. The main surgical targets are the ventromedial nucleus, the paraventricular nucleus, and the arcuate nucleus of the hypothalamus. Further, other surgical targets are found, such as surgical removal of brown adipose tissue or ovariectomy.

## 4. Models involving *Helicobacter pylori* and the gastrointestinal gut-related diseases

*Helicobacter pylori (H. pylori)* is a bacterium that infects the stomach and despite the very low pH of this microenvironment, that strain resides in the organ. Several studies have demonstrated that this infection usually happens during childhood; however, it is still unclear how the mechanism of transmission occurs [99]. This bacterium is a very common cause of peptic ulcers and gastritis, and it is considered a grade 1 carcinogen by the World Health Organization.

The global prevalence of *H. pylori* has reduced from 52.6% before 1990 to 43.9% in adults during 2015–2022, and 35.1% in kids and adolescents during 2015–2022 [100]. A high percentage of infected people are asymptomatic for *H. pylori* infection. The frequent signs and symptoms of gastritis and peptic ulcer need investigation and testing for infection.

The treatment of the bacteria has become increasingly difficult to achieve due to the emergence of strains resistant to conventional antimicrobials [101]. Reinforcing the increases of antimicrobial resistance, Benigno et al., in a recent study, evaluated 198 patients, consisting of 153 gastritis, 24 gastric cancer, 21 peptic ulcer, and 24 gastric juices from asymptomatic patients.

All the individuals were diagnosed positive for *H. pylori* infection by string test and 23S rRNA. Clarithromycin was assessed by quantitative real-time PCR for A2142G and/or A2143G point of mutations. The resistance prevalence against clarithromycin was 14.4% (32/222), and the mutation was less frequent in cagA+ (11.4%) in comparison to cagA- (23.6%) [102].

Studies using murine models of gastric cancer and other gastrointestinal diseases associated with *H. pylori* infection are essential for identifying virulence factors and new forms of treatment. Current treatment for infection consists of eradicating the bacteria consisting in triple and quadruple therapy typically involving a proton-pump inhibitor (PPI) and two antibiotics (clarithromycin and amoxicillin or metronidazole), and PPI, bismuth, metronidazole, and tetracycline [103].

*H. pylori* is classified as a flagellated Gram-negative bacillus and has virulence mechanisms that favor colonization of the stomach. Virulence mechanisms are expressed through the presence of genes that ensure the production of proteins essential for the maintenance of the bacterium in the environment. Some strains of *H. pylori* have genes associated with aggressive virulence mechanisms, such as genes found in the cytotoxin-associated gene (CAG) pathogenicity island. This region of the bacterial genetic material has genes responsible for the expression of virulent characteristics of the bacteria, such as cagA, vacA, and virB11 genes [104–106].

In addition, the bacteria have important mechanisms of adhesion to the stomach wall, through outer membrane proteins, expressed through the BabA, iceA, and oipA genes. These proteins allow adhesion to the Lewis-b antigen, present in gastric epithelial cells and become consistently expressed in contact with this epithelium [106].

The main *Helicobacter* species that are pathogenic to humans are *Helicobacter pylori*, *H. felis,* and *H. suis*. They are used in studies with rodents to induce carcinogenesis, especially in C57BL/6, C3H, and B6129 mice. Carcinogenesis induction in these mice lasts approximately 15 months for C57BL/6, 13 months for C3H, and 8 months for B6129 [101, 107].

Beyond the introduction of the bacteria, some mechanisms can be adopted to accelerate the carcinogenesis process induced by *H. pylori*, such as adherence to a high-salt diet; utilization of carcinogenic compounds, such as N-methyl-N-nitrosourea (MNU) and methylnitronitrosoguanidine; and the use of knockout models [108]. The genetic modifications are the most widely used knockout models as highlighted in the insulin-gastrin (INS-GAS) model, interferon-gamma (IFN-$\gamma$) model, tumor necrosis factor-alpha (TNF-$\alpha$) model, interleukin (IL)-10, and p27-deficient mice [101] and also knocking out of antioxidant protective genes, such as the mutR homolog-1 (Mth1)-deficient mouse or the trefoil factor 1 (Tff1)-deficient mouse [107].

In addition, the INS-GAS genetic model spontaneously develops carcinogenesis, a process that is accelerated by inflammation mechanisms initiated by *H. pylori* infection. The main transgenic models found of them are associated with the bacteria and are described as modifications in the cytotoxin-associated gene A (CagA)-transgenic *H. pylori* and IL-1$\beta$ transgenic mice. In addition, models that express protooncogenes, such as the CEA/SV40T mouse, are also used for carcinogenesis development [101].

Another gastrointestinal condition, with very relevant prevalence is gastric cancer. This cancer is the fifth most common and the third leading cause of cancer death worldwide, according to the Global Cancer Observatory (GCO) [109]. Gastric cancer is common and has a poor prognosis due to late diagnosis. In addition, current treatment options for gastric cancer are limited.

The development of gastric cancer is related to genetic, epigenetic, and infectious factors. Gastrointestinal infections caused by virulent strains of *Helicobacter pylori* are found in most cases of gastric cancer [110].

As mentioned before, *H. pylori* has several genes recognized as virulence factors. These virulence factors are related to the development of gastric cancer through a bacterium's secretion system that transports messenger molecules that activate human oncogenes, such as the YAP gene, or inactivate cell cycle regulatory proteins, such as p53, p16INK4a, c-Myc, Bcl-2, and Bax, which can result in the development of gastric cancer [111, 112].

In this regard, studies on the influence of *H. pylori* on the pathogenesis of gastric cancer using animal models allow *in vivo* visualization of the infection's pathophysiological mechanisms, immune alterations, and neoplasia development.

Furthermore, studying the development of malignant neoplasms in the gastrointestinal tract using animal models is essential for identifying the initial stages of the disease, as well as the application of drugs and technologies under development. The main malignant neoplasm associated with IBD is colorectal cancer, which ranks as the third most common cancer in the world and has projections for the emergence of 3.2 million new cases by 2040 [113].

The main murine models used in laboratory research are mice of the *Mus musculus* species and rats of the *Rattus norvegicus* species [68]. The animal models used in the studies of neoplasms may present the suppression of tumor suppressor genes, such as the TP53 gene, which configures them as knockout animal models. Other models may present a mutation in the APC gene, configuring knockin animal models [114].

Animal models have been widely used to study the mechanisms of gastric carcinogenesis. Several models are used to induce this carcinogenesis in animals, and among these models, those that use chemical agents, bacterial infections, genetic alterations, and xenotransplantation to induce random and targeted genetic mutations, respectively, are crucial tools in biomedical research, used to study the pathogenesis, progression and treatment of this disease [115].

These models are fundamental to understanding the molecular and cellular mechanisms underlying gastric cancer, which is one of the most common and lethal types of cancer in the world. Thus, the use of animal models has provided a valuable platform for understanding the molecular mechanisms, drug discovery, and validation, as well as for the development of diagnostic and prognostic biomarkers, for early detection and monitoring of disease evolution/regression and for defining more specific treatment protocols for each patient.

Predominantly chemical and bacterial induction are the most used. The best model to chemically induce gastric carcinoma in animals involves the administration of MNU, an N-nitroso compound generated by anaerobic intestinal bacteria after ingestion of nitrates and nitrites [116]. This model has demonstrated the importance of several signaling pathways in gastric tumorigenesis. However, the precise mechanism of MNU-induced carcinogenesis is not completely understood. In models of bacterial induction of adenocarcinoma, *Helicobacter* infection (*H. felis* and *H. pylori* SS1 Sydney) is used. *Helicobacter* infection models have been crucial for studying immune responses during chronic infection and gastritis, as well as the influence of factors such as diet, microbiota, and gender on disease progression [116].

Despite their value, animal models of gastric cancer have limitations. Differences between animal and human biology may affect the clinical relevance of results. Furthermore, the heterogeneity of human gastric cancer is not always faithfully replicated in animal models, which may limit the translation of findings.

The importance of carefully matching the choice of gastric cancer induction model in the animal with the specific biological question being investigated ensures relevant outcomes.

Animal models of gastric cancer are essential tools for research and development of new treatments, despite the limitations. They continue to be an area of intense investigation, aiming to bring models with closer characteristics to the complexity of human disease presentation.

## 5. CNS-related models and impact on gastrointestinal tract

There are many CNS diseases that have an impact on gastrointestinal tract, which was found by recent studies. Anxiety and depression are two major diseases that the studies have established connections with and the evidence has shown the negative impact in both systems, CNS and GI. From gut to brain many biochemical, hormonal or immunoinflammatory pathways could be affected by any dysregulation of the gastrointestinal system.

Modeling complex human disorders like anxiety and depression in animals is a challenging task due to the variability in symptoms and the intricate biomolecular system involved. Current animal models of depression are based on criteria that the model must produce a similar behavioral and comparable neurobiological phenotype and respond to clinically effective antidepressant treatments. However, these models have limitations and may not fully capture the diversity of biological systems and behavioral symptoms observed in anxiety and depressive patients [117, 118].

Studies using murine models of stress require a variety of measures to assess stress-induced behaviors that mimic clinical depression, including preference for a sweet substance to characterize anhedonia (similar to the inability to feel pleasure in humans), using spatial memory tasks to assess cognitive impairment, monitoring grooming behavior to gauge feelings of worthlessness, and measuring the duration of social interactions to determine dysfunctional social behavior [118–120].

Additionally, anxiety-related behaviors and changes in locomotor activity are often assessed using open field or elevated plus maze tests. Behavioral despair or passive coping is commonly measured through forced swimming or tail suspension tests, although it remains unclear how these tests translate to human depression symptoms. Recently, the ethics and validity of the tail suspension and forced swimming tests, which measure behavioral despair, have been questioned, highlighting the need for novel behavioral tests that more accurately reflect the human condition [121].

Several years have been spent on researching mental disorders such as anxiety and depression and there is still an enormous gap on the knowledge about these conditions. Thus the animal models can address several mechanisms involving stress/depression behavior. One of the methods, which usually referred to an early life stress, is based on inducing stress on mice by the Maternal Separation (MS). This method is based on separation of pups from their mothers during progressive scale of time (hours a day for several days) leading to a dysregulation on hypothalamic-pituitary-adrenal (HPA) axis. It is demonstrated that the disarrangements on this axis could be related to depression [122].

In CNS, there are several modifications associated to this model, and major modifications are the inflammatory pattern increases and elevating the circulating levels of well-know inflammatory cytokines IL-1β, IL-6, and TNF-α in pups under the MS protocol [123]. Also, MS is used to model the IBS, and the model induced several dysfunctions in GI that increased both visceral sensitivity [124] and vulnerability to colitis.

Also, this model is associated with several modifications in intestinal microbiota. Enqi et al., studying the effects of IBS, found increased levels of *Clostridium IV, Corynebacterium, Rothia, Elusimicrobium, Romboutsia, Allobaculum,* and

*Parasutterella*, and their related taxa were specifically associated with MS group. This study highlighted how the insult of protocol can change for a deleterious microbiota in animals [125].

Chronic restraint stress (CRS) is another murine model of stress that frequently is associated with gastrointestinal negative impact. The model consists in restraining animals daily for hours over a period of 2 or 3 weeks [122]. CNS outcomes from this model are frequently atrophic pyramidal cells, corticosterone elevation levels in hippocampus. Moreover, in the gut, this model shows an increase in the permeability of the gut epithelia suggesting the breaking of the gut barrier [126], and also the restrain stress model can change the microbiota in gut according to studies from several authors.

The study conducted by Deng et al. demonstrated that animals subjects of the restrain stress model exhibited several behavior changes such as depression disorders, and the authors suggested that these modifications were associated to the changes in microbiota modifications due to stress model [127].

In addition, chronic unpredictable mild stress (CUMS) is composed by several mild stressor mechanisms including restraint, water and food deprivation, overcrowding, cage tilt, and chronic social defeat stress. CUMS is a widely used method by dysregulating the HPA axis affecting brain regions involved in depression such as the hippocampus and prefrontal cortex [122].

The chronic social defeat stress (CSDS) is a model where young males are put together with an older dominant mice aiming to erupt a social conflict to develop anhedonia, anxiety like and depressive behaviors (e.g., helplessness behavior), and also the administration of glucocorticoid could lead to a dysregulation of HPA axis functioning and mimics the anxiety-depression behavior [51, 128].

Thus, gastrointestinal diseases are a very large group of diseases and further chapters must address the variability beyond the animal models, in addition to the in vitro model.

## 6. Conclusions

Thus, this chapter summarizes the main animal methods used in several gastrointestinal disorders highlighting the main advantages and describing the disadvantages for each of them. The best choice for one of them depends on the question that needs to be addressed, the costs involved in the project, and the level of personnel.

## Acknowledgements

The authors want to acknowledge the Instituto de Educação Médica – IDOMED and the FIOCRUZ-PASTEUR Institute, for the financial support for the publication of this chapter.

## Author contribution

All the authors have contributed equally to the chapter. FMF: conceptualization, writing—original draft, and review and Editing. BG and GF: writing—original draft;

RBM and CPMF: conceptualization writing—original draft. OGRA: writing—original draft, conceptualization, writing—review and editing, and supervision.

## Conflict of interest

The authors declare no conflict of interest.

## Author details

Fabiano Marcelo Fabris[1], Beatriz Gomes Pinto[1], Geraldo de França Júnior[1], Radamés Bezerra Melo[2], Clarissa Perdigão Melo Ferraz[3] and Orleâncio Gomes Ripardo de Azevedo[1*]

1 Faculdade Estacio de Canindé (Instituto de Educação Médica – IDOMED), School of Medicine, Canindé, Brazil

2 Paulo Picanço School of Dentistry, Fortaleza, Ceara, Brazil

3 Center Pasteur-Fiocruz of Immunology and Immunotherapy, Eusébio, Ceara, Brazil

*Address all correspondence to: orleancio@gmail.com

## References

[1] Accarie A, Vanuytsel T. Animal models for functional gastrointestinal disorders. Frontiers in Psychiatry. 2020;**11**:1-23

[2] Woodward B. Fidelity in animal modeling: Prerequisite for a mechanistic research front relevant to the inflammatory incompetence of acute pediatric malnutrition. International Journal of Molecular Sciences. 2016;**17**(4):541

[3] Attia S, Feenstra M, Swain N, Cuesta M, Bandsma RHJ. Starved guts. Journal of Pediatric Gastroenterology and Nutrition. 2017;**65**(5):491-495

[4] Ueno PM, Oriá RB, Maier EA, Guedes M, de Azevedo OG, Wu D, et al. Alanyl-glutamine promotes intestinal epithelial cell homeostasis in vitro and in a murine model of weanling undernutrition. American Journal of Physiology. Gastrointestinal and Liver Physiology. 2011;**301**(4):G612-G622

[5] Salameh E, Morel FB, Zeilani M, Déchelotte P, Marion-Letellier R. Animal models of undernutrition and enteropathy as tools for assessment of nutritional intervention. Nutrients. 2019;**11**(9):2233

[6] Massironi S, Viganò C, Palermo A, Pirola L, Mulinacci G, Allocca M, et al. Inflammation and malnutrition in inflammatory bowel disease. The Lancet Gastroenterology & Hepatology. 2023;**8**(6):579-590

[7] Ladd FVL, Ladd AABL, Ribeiro AACM, Costa SBC, Coutinho BP, Feitosa GAS, et al. Zinc and glutamine improve brain development in suckling mice subjected to early postnatal malnutrition. Nutrition. 2010;**26**(6):662-670

[8] Wen C, Chen D, Zhong R, Peng X. Animal models of inflammatory bowel disease: Category and evaluation indexes. Gastroenterology Report (Oxford). 2023;**12**:1-13

[9] Selvaratnam S, Gullino S, Shim L, Lee E, Lee A, Paramsothy S, et al. Epidemiology of inflammatory bowel disease in South America: A systematic review. World Journal of Gastroenterology. 2019;**25**(47):6866-6875

[10] Sands BE. From symptom to diagnosis: Clinical distinctions among various forms of intestinal inflammation. Gastroenterology. 2004;**126**(6):1518-1532

[11] Saeid Seyedian S, Nokhostin F, Dargahi MM. A review of the diagnosis, prevention, and treatment methods of inflammatory bowel disease. Journal of Medicine and Life. 2019;**12**(2):113-122

[12] Shen B. Interventional inflammatory bowel disease: Endoscopic therapy of complications of Crohn's disease. Gastroenterology Report (Oxford). 2022;**10**:1-10

[13] Yu LCH. Microbiota dysbiosis and barrier dysfunction in inflammatory bowel disease and colorectal cancers: Exploring a common ground hypothesis. Journal of Biomedical Science. 2018;**25**(1):79

[14] Parente JML. Inflammatory bowel disease in an underdeveloped region of Northeastern Brazil. World Journal of Gastroenterology. 2015;**21**(4):1197

[15] Alavinejad P, Hashemi SJ, Behl N, Hormati A, Elbasuny A, Daryani NE, et al. Inflammatory bowel disease evolution in the past two decades: A chronological multinational study. EClinicalMedicine. 2024;**70**:102542

[16] Adams SM, Close ED, Shreenath AP. Ulcerative colitis: Rapid evidence review. American Family Physician. 2022;**105**(4):406-411

[17] Jostins L, Ripke S, Weersma RK, Duerr RH, McGovern DP, Hui KY, et al. Host–microbe interactions have shaped the genetic architecture of inflammatory bowel disease. Nature. 2012;**491**(7422):119-124

[18] Hodge C, Taylor C. Vitamin A Deficiency. Treasure Island, FL: StatPearls Publishing; Jan 2024

[19] Sit B, Fakoya B, Waldor MK. Animal models for dissecting vibrio cholerae intestinal pathogenesis and immunity. Current Opinion in Microbiology. 2022;**65**:1-7

[20] Zhan X, Wang F, Bi Y, Ji B. Animal models of gastrointestinal and liver diseases. Animal models of acute and chronic pancreatitis. American Journal of Physiology-Gastrointestinal and Liver Physiology. 2016;**311**(3):G343-G355

[21] Yoo J, Groer M, Dutra S, Sarkar A, McSkimming D. Gut microbiota and immune system interactions. Microorganisms. 2020;**8**(10):1587

[22] Allaire JM, Crowley SM, Law HT, Chang SY, Ko HJ, Vallance BA. The intestinal epithelium: Central coordinator of mucosal immunity. Trends in Immunology. 2018;**39**(9):677-696

[23] Chassaing B, Aitken JD, Malleshappa M, Vijay-Kumar M. Dextran Sulfate sodium (DSS)-induced colitis in mice. Current Protocols in Immunology. 2014;**104**(1):1-16

[24] Baydi Z, Limami Y, Khalki L, Zaid N, Naya A, Mtairag EM, et al. An update of research animal models of inflammatory bowel disease. The Scientific World Journal. 2021;**2021**:1-12

[25] Owusu G, Obiri DD, Ainooson GK, Osafo N, Antwi AO, Duduyemi BM, et al. Acetic acid-induced ulcerative colitis in Sprague Dawley rats is suppressed by Hydroethanolic extract of Cordia vignei leaves through reduced serum levels of TNF-α and IL-6. International Journal of Chronic Diseases. 2020;**2020**:1-11

[26] Fox JG, Ge Z, Whary MT, Erdman SE, Horwitz BH. Helicobacter hepaticus infection in mice: Models for understanding lower bowel inflammation and cancer. Mucosal Immunology. 2011;**4**(1):22-30

[27] Sharma M, Sharma S, Wadhwa J. Improved uptake and therapeutic intervention of curcumin via designing binary lipid nanoparticulate formulation for oral delivery in inflammatory bowel disorder. Artificial Cells, Nanomedicine, and Biotechnology. 2019;**47**(1):45-55

[28] Takahashi N, Kitazawa C, Itani Y, Awaga Y, Hama A, Hayashi I, et al. Exploratory clinical characterization of experimentally-induced ulcerative colitis nonhuman primates. Heliyon. 2020;**6**(1):e03178

[29] Morales Fénero C, Amaral MA, Xavier IK, Padovani BN, Paredes LC, Takiishi T, et al. Short chain fatty acids (SCFAs) improves TNBS-induced colitis in zebrafish. Current Research in Immunology. 2021;**2**:142-154

[30] Sestak K, Merritt CK, Borda J, Saylor E, Schwamberger SR, Cogswell F, et al. Infectious agent and immune response characteristics of chronic Enterocolitis in captive rhesus macaques. Infection and Immunity. 2003;**71**(7):4079-4086

[31] Ma S, Yeom J, Lim YH. Dairy Propionibacterium freudenreichii

ameliorates acute colitis by stimulating MUC2 expression in intestinal goblet cell in a DSS-induced colitis rat model. Scientific Reports. 2020;**10**(1):5523

[32] de Brito TV, Júnior GJD, da Cruz Júnior JS, Silva RO, da Silva Monteiro CE, Franco AX, et al. Gabapentin attenuates intestinal inflammation: Role of PPAR-gamma receptor. European Journal of Pharmacology. 2020;**873**:172974

[33] Costa-Filho HB, Sales TMAL, Paula SM, Nicolau LAD, Queiroga ML, Havt A, et al. Role of cyclooxygenases 1 and 2 in the maintenance of colonic mucosal integrity in an experimental colitis model. Brazilian Journal of Medical and Biological Research. 2023;**56**:1-10

[34] Brito TV, FCN B, Silva RO, Dias Júnior GJ, Júnior JSC, Franco ÁX, et al. Sulfated polysaccharide from the marine algae Hypnea musciformis inhibits TNBS-induced intestinal damage in rats. Carbohydrate Polymers. 2016;**151**:957-964

[35] Algieri F, Rodriguez-Nogales A, Garrido-Mesa J, Camuesco D, Vezza T, Garrido-Mesa N, et al. Intestinal anti-inflammatory activity of calcium pyruvate in the TNBS model of rat colitis: Comparison with ethyl pyruvate. Biochemical Pharmacology. 2016;**103**:53-63

[36] Dutra NLS, de Brito TV, de Magalhães DA, Sousa SG, Batista JA, Pereira CMC, et al. Sulfated polysaccharide extracted from seaweed Gracilaria caudata attenuates acetic acid-induced ulcerative colitis. Food Hydrocolloids. 2021;**111**:106221

[37] Anderson CA, Boucher G, Lees CW, Franke A, D'Amato M, Taylor KD, et al. Meta-analysis identifies 29 additional ulcerative colitis risk loci, increasing the number of confirmed associations to 47. Nature Genetics. 2011;**43**(3):246-252

[38] Khor B, Gardet A, Xavier RJ. Genetics and pathogenesis of inflammatory bowel disease. Nature. 2011;**474**(7351):307-317

[39] Pizarro TT, Pastorelli L, Bamias G, Garg RR, Reuter BK, Mercado JR, et al. SAMP1/YitFc mouse strain: A spontaneous model of Crohn's disease-like ileitis. Inflammatory Bowel Diseases. 2011;**17**(12):2566-2584

[40] Ahern PP, Schiering C, Buonocore S, McGeachy MJ, Cua DJ, Maloy KJ, et al. Interleukin-23 drives intestinal inflammation through direct activity on T cells. Immunity. 2010;**33**(2):279-288

[41] Mizoguchi E, Low D, Ezaki Y, Okada T. Recent updates on the basic mechanisms and pathogenesis of inflammatory bowel diseases in experimental animal models. Intestinal Research. 2020;**18**(2):151-167

[42] McLean MH, Andrews C, Hanson ML, Baseler WA, Anver MR, Senkevitch E, et al. Interleukin-27 is a potential rescue therapy for acute severe colitis through Interleukin-10–dependent, T-cell–independent attenuation of colonic mucosal innate immune responses. Inflammatory Bowel Diseases. 2017;**23**(11):1983-1995

[43] Nakamura Y, Igaki K, Komoike Y, Yokoyama K, Tsuchimori N. Malt1 inactivation attenuates experimental colitis through the regulation of Th17 and Th1/17 cells. Inflammation Research. 2019;**68**(3):223-230

[44] Mizoguchi E, Subramaniam R, Okada T, Mizoguchi A. A review of selected IBD biomarkers: From animal models to bedside. Diagnostics. 2021;**11**(2):207

[45] Abraham C, Cho JH. Inflammatory bowel disease. New England Journal of Medicine. 2009;**361**(21):2066-2078

[46] Xing T, Camacho Salazar R, Chen YH. Animal models for studying epithelial barriers in neonatal necrotizing enterocolitis, inflammatory bowel disease and colorectal cancer. Tissue Barriers. 2017;**5**(4):e1356901

[47] Katsandegwaza B, Horsnell W, Smith K. Inflammatory bowel disease: A review of pre-clinical murine models of human disease. International Journal of Molecular Sciences. 2022;**23**(16):9344

[48] Vieujean S, Caron B, Haghnejad V, Jouzeau JY, Netter P, Heba AC, et al. Impact of the Exposome on the Epigenome in inflammatory bowel disease patients and animal models. International Journal of Molecular Sciences. 2022;**23**(14):7611

[49] Barbalho SM, de Goulart RA, dos Batista GLSA. Vitamin A and inflammatory bowel diseases: From cellular studies and animal models to human disease. Expert Review of Gastroenterology & Hepatology. 2019;**13**(1):25-35

[50] Rangan P, Choi I, Wei M, Navarrete G, Guen E, Brandhorst S, et al. Fasting-mimicking diet modulates microbiota and promotes intestinal regeneration to reduce inflammatory bowel disease pathology. Cell Reports. 2019;**26**(10):2704-2719.e6

[51] Liu S, Chang J, Long N, Beckwith K, Talhouarne G, Brooks JJ, et al. Endogenous CRF in rat large intestine mediates motor and secretory responses to stress. Neurogastroenterology & Motility. 2016;**28**(2):281-291

[52] Britton GJ, Contijoch EJ, Mogno I, Vennaro OH, Llewellyn SR, Ng R, et al. Microbiotas from humans with inflammatory bowel disease Alter the balance of gut Th17 and RORγt+ regulatory T cells and exacerbate colitis in mice. Immunity. 2019;**50**(1):212-224.e4

[53] WHO. The Top 10 Causes of Death. 2024. Available from: https://www.who.int/news-room/fact-sheets/detail/the-top-10-causes-of-death [Accessed: August, 2024]

[54] Murray PR, Rosenthal KS, Pfaller MA. Medical Microbiology. 9th ed. Vol. 30. Philadelphia: Elsevier; 2020. pp. 1370-2111

[55] Wei L, Singh R, Ro S, Ghoshal UC. Gut microbiota dysbiosis in functional gastrointestinal disorders: Underpinning the symptoms and pathophysiology. JGH Open. 2021;**5**(9):976-987

[56] Hwang IY, Koh E, Wong A, March JC, Bentley WE, Lee YS, et al. Engineered probiotic *Escherichia coli* can eliminate and prevent Pseudomonas aeruginosa gut infection in animal models. Nature Communications. 2017;**8**(1):15028

[57] Azevedo OGR, Bolick DT, Roche JK, Pinkerton RF, Lima AAM, Vitek MP, et al. Apolipoprotein E plays a key role against Cryptosporidial infection in transgenic undernourished mice. PLoS One. 2014;**9**(2):e89562

[58] Skerniskyte J, Mulet C, André AC, Anderson MC, Injarabian L, Buck A, et al. Ascorbate deficiency increases progression of shigellosis in Guinea pigs and mice infection models. Gut Microbes. 2023;**15**(2):1-18

[59] World Health Organization. Malnutrition. 2024. Available from: https://www.who.int/news-room/fact-sheets/detail/malnutrition [Accessed: August, 2024]

[60] Hidalgo-Villeda F, Million M, Defoort C, Vannier T, Svilar L, Lagier M, et al. Prolonged dysbiosis and altered immunity under nutritional intervention in a physiological mouse model of severe acute malnutrition. iScience. 2023;**26**(6):106910

[61] M. Elia BA for P and ENAG on M. Front Cover Image for Guidelines for Detection and Management of Malnutrition Guidelines for Detection and Management of Malnutrition. Maidenhead: BAPEN; 2000; 2000

[62] Gulland A. Malnutrition and obesity coexist in many countries, report finds. BMJ. 2016:i3351

[63] Bouillon R, Marcocci C, Carmeliet G, Bikle D, White JH, Dawson-Hughes B, et al. Skeletal and extraskeletal actions of vitamin D: Current evidence and outstanding questions. Endocrine Reviews. 2019;**40**(4):1109-1151

[64] Kramarz C, Murphy E, Reilly MM, Rossor AM. Nutritional peripheral neuropathies. Journal of Neurology, Neurosurgery, and Psychiatry. 2024;**95**(1):61-72

[65] Pasricha SR, Tye-Din J, Muckenthaler MU, Swinkels DW. Iron deficiency. The Lancet. 2021;**397**(10270):233-248

[66] Hatch-McChesney A, Lieberman HR. Iodine and iodine deficiency: A comprehensive review of a re-emerging issue. Nutrients. 2022;**14**(17):3474

[67] Shlisky J, Mandlik R, Askari S, Abrams S, Belizan JM, Bourassa MW, et al. Calcium deficiency worldwide: Prevalence of inadequate intakes and associated health outcomes. Annals of the New York Academy of Sciences. 2022;**1512**(1):10-28

[68] Doulberis M, Papaefthymiou A, Polyzos SA, Katsinelos P, Grigoriadis N, Srivastava DS, et al. Rodent models of obesity. Minerva Endocrinologica. 2020;**45**:3

[69] de Oliveira CAM, Latorraca MQ, de Mello MAR, Carneiro EM. Mechanisms of insulin secretion in malnutrition: Modulation by amino acids in rodent models. Amino Acids. 2011;**40**(4):1027-1034

[70] Pacheco-Martínez MM, Cortés-Barberena E, Cervantes-Ríos E, del Carmen G-RM, Rodríguez-Cruz L, Ortiz-Muñiz R. Moderate malnutrition in rats induces somatic gene mutations. Mutation Research/Fundamental and Molecular Mechanisms of Mutagenesis. 2016;**789**:26-32

[71] Oriá RB, Patrick PD, Oriá MOB, Lorntz B, Thompson MR, Azevedo OGR, et al. ApoE polymorphisms and diarrheal outcomes in Brazilian shanty town children. Brazilian Journal of Medical and Biological Research. 2010;**43**(3):249-256

[72] Mitter SS, Oriá RB, Kvalsund MP, Pamplona P, Joventino ES, Mota RMS, et al. Apolipoprotein E4 influences growth and cognitive responses to micronutrient supplementation in shantytown children from Northeast Brazil. Clinics. 2012;**67**(1):11-18

[73] Oriá RB, Costa DVS, de Medeiros PHQS, Roque CR, Dias RP, Warren CA, et al. Myeloperoxidase as a biomarker for intestinal-brain axis dysfunction induced by malnutrition and cryptosporidium infection in weanling mice. The Brazilian Journal of Infectious Diseases. 2023;**27**(3):102776

[74] Ribeiro SA, de Rodrigues FAP, de Clementino MAF, do Veras HN, RCL S, de Medeiros PHQS, et al.

Consumption of a multi-deficient diet causes dynamic changes in the intestinal morphofunctional barrier, body composition and impaired physical development in post-weaning mice. British Journal of Nutrition. 2023;**129**(5):745-758

[75] Santos MJS, Canuto KM, de Aquino CC, Martins CS, Brito GAC, Pessoa TMRP, et al. A Brazilian regional basic diet-induced chronic malnutrition drives liver inflammation with higher ApoA-I activity in C57BL6J mice. Brazilian Journal of Medical and Biological Research. 2020;**53**(6)

[76] Freitas RS, Roque CR, Matos GA, Belayev L, de Azevedo OGR, Alvarez-Leite JI, et al. Immunoinflammatory role of apolipoprotein E4 in malnutrition and enteric infections and the increased risk for chronic diseases under adverse environments. Nutrition Reviews. 2022;**80**(5):1001-1012

[77] Siegfried Z, Berry EM, Hao S, Avraham Y. Animal models in the investigation of anorexia. Physiology & Behavior. 2003;**79**(1):39-45

[78] Herselman MF, Bailey S, Bobrovskaya L. The effects of stress and diet on the "brain–gut" and "gut–brain" pathways in animal models of stress and depression. International Journal of Molecular Sciences. 2022;**23**(4):2013

[79] Silva Neto LB, da, Chaves Filho AJM, Casadevall MQFC, Azevedo OGR de, Macêdo DS, Vasconcelos PRL de. Ad libitum consumption of milk supplemented with omega 3, 6, and 9 oils from infancy to middle age alters behavioral and oxidative outcomes in male mice. Brazilian Journal of Medical and Biological Research. 2022;**55**

[80] World Health Organization - WHO. Obesity and Overweight. 2024

[81] Joyner MJ. Rethinking animal models and human obesity. Physiology. 2014;**29**(6):384-385

[82] Goodarzi MO. Genetics of obesity: What genetic association studies have taught us about the biology of obesity and its complications. The Lancet Diabetes and Endocrinology. 2018;**6**(3):223-236

[83] Kleinert M, Clemmensen C, Hofmann SM, Moore MC, Renner S, Woods SC, et al. Animal models of obesity and diabetes mellitus. Nature Reviews. Endocrinology. 2018;**14**(3):140-162

[84] Liu R, Hong J, Xu X, Feng Q, Zhang D, Gu Y, et al. Gut microbiome and serum metabolome alterations in obesity and after weight-loss intervention. Nature Medicine. 2017;**23**(7):859-868

[85] Fulton S, Décarie-Spain L, Fioramonti X, Guiard B, Nakajima S. The menace of obesity to depression and anxiety prevalence. Trends in Endocrinology & Metabolism. 2022;**33**(1):18-35

[86] Ghanemi A, Yoshioka M, St-Amand J. Obese animals as models for numerous diseases: Advantages and applications. Medicina (B Aires). 2021;**57**(5):399

[87] Doyle A, McGarry MP, Lee NA, Lee JJ. The construction of transgenic and gene knockout/knockin mouse models of human disease. Transgenic Research. 2012;**21**(2):327-349

[88] Inoue K, ichi, Toyoda S, Jojima T, Abe S, Sakuma M, Inoue T. Time-restricted feeding prevents high-fat and high-cholesterol diet-induced obesity but fails to ameliorate atherosclerosis in apolipoprotein

E-knockout mice. Experimental Animals. 2021;**70**(2):194-202

[89] Ohno T, Miyasaka Y, Yoshida K, Kobayashi M, Horio F, Yokoi N, et al. A novel model mouse for type 2 diabetes mellitus with early onset and persistent hyperglycemia. Experimental Animals. 2022;**71**(4):22-0061

[90] Li J, Wu H, Liu Y, Yang L. High fat diet induced obesity model using four strains of mice: Kunming, C57BL/6, BALB/c and ICR. Experimental Animals. 2020;**69**(3):326-335

[91] Sanapalli BKR, Yele V, Singh MK, Thaggikuppe Krishnamurthy P, Karri VVSR. Preclinical models of diabetic wound healing: A critical review. Biomedicine & Pharmacotherapy. 2021;**142**:111946

[92] Srinivasan K, Ramarao P. Animal models in type 2 diabetes research: An overview. The Indian Journal of Medical Research. 2007;**125**(3):451-472

[93] Lee JH, Yang SH, Oh JM, Lee MG. Pharmacokinetics of drugs in rats with diabetes mellitus induced by alloxan or streptozocin: Comparison with those in patients with type I diabetes mellitus. Journal of Pharmacy and Pharmacology. 2010;**62**(1):1-23

[94] He Y, Wu N, Yu W, Li L, OuYang H, Liu X, et al. Research progress on the experimental animal model of gestational diabetes mellitus. Diabetes, Metabolic Syndrome and Obesity. 2020;**13**:4235-4247

[95] Bastías-Pérez M, Zagmutt S, Soler-Vázquez MC, Serra D, Mera P, Herrero L. Impact of adaptive thermogenesis in mice on the treatment of obesity. Cells. 2020;**9**(2):316

[96] Tokunaga K, Matsuzawa Y, Fujioka S, Kobatake T, Keno Y, Odaka H, et al. PVN-lesioned obese rats maintain ambulatory activity and its circadian rhythm. Brain Research Bulletin. 1991;**26**(3):393-396

[97] Deng X, Feng X, Li S, Gao Y, Yu B, Li G. Influence of the hypothalamic paraventricular nucleus (PVN) on heart rate variability (HRV) in rat hearts via electronic lesion. Bio-medical Materials and Engineering. 2015;**26**(s1):S487-S495

[98] Secher A, Jelsing J, Baquero AF, Hecksher-Sørensen J, Cowley MA, Dalbøge LS, et al. The arcuate nucleus mediates GLP-1 receptor agonist liraglutide-dependent weight loss. Journal of Clinical Investigation. 2014;**124**(10):4473-4488

[99] Kayali S, Manfredi M, Gaiani F, Bianchi L, Bizzarri B, Leandro G, et al. *Helicobacter pylori*, transmission routes and recurrence of infection: State of the art. Acta Bio-Medica. 2018;**89**(8-S):72-76

[100] Chen YC, Malfertheiner P, Yu HT, Kuo CL, Chang YY, Meng FT, et al. Global prevalence of *Helicobacter pylori* infection and incidence of gastric cancer between 1980 and 2022. Gastroenterology. 2024;**166**(4):605-619

[101] Ansari S, Yamaoka Y. Animal models and *Helicobacter pylori* infection. Journal of Clinical Medicine. 2022;**11**(11):3141

[102] da Benigno TGS, Ribeiro Junior HL, de Azevedo OGR, Pinheiro RF, de Oliveira RTG, Maciel FS, et al. Clarithromycin-resistant *H. Pylori* primary strains and virulence genotypes in the Northeastern region of Brazil. Revista do Instituto de Medicina Tropical de São Paulo. 2022:64

[103] Lee YC, Dore MP, Graham DY. Diagnosis and treatment of *Helicobacter*

*pylori* infection. Annual Review of Medicine. 2022;**73**(1):183-195

[104] Lima VP, Rabenhorst SHB. Genes associados à virulência de *Helicobacter pylori*. Revista Brasileira de Cancerologia. 2009;**55**(4):389-396

[105] Lima VP, de Silva-Fernandes IJL, MKS A, SHB R. Prevalence of *Helicobacter pylori* genotypes (vacA, cagA, cagE and virB11) in gastric cancer in Brazilian's patients: An association with histopathological parameters. Cancer Epidemiology. 2011;**35**(5):e32-e37

[106] Lima VP, de Lima MAP, Ferreira MVP, Barros MAP, Rabenhorst SHB. The relationship between *Helicobacter pylori* genes cagE and virB11 and gastric cancer. International Journal of Infectious Diseases. 2010;**14**(7):e613-e617

[107] Burkitt MD, Duckworth CA, Williams JM, Pritchard DM. *Helicobacter pylori*-induced gastric pathology: Insights from in vivo and ex vivo models. Disease Models & Mechanisms. 2017;**10**(2):89-104

[108] Kabir S. Effect of *Helicobacter pylori* eradication on incidence of gastric cancer in human and animal models: Underlying biochemical and molecular events. Helicobacter. 2009;**14**(3):159-171

[109] Sung H, Ferlay J, Siegel RL, Laversanne M, Soerjomataram I, Jemal A, et al. Global cancer statistics 2020: GLOBOCAN estimates of incidence and mortality worldwide for 36 cancers in 185 countries. CA: A Cancer Journal for Clinicians. 2021;**71**(3):209-249

[110] IARC. Schistosomes, liver flukes and *Helicobacter pylori*. In: Monographs on the Identification of Carcinogenic Hazards to Humans. Vol. 61. Lyon: IARC; 1994

[111] Braga LLBC, Batista MHR, De Azevedo OGR, Da Silva Costa KC, Gomes AD, Rocha GA, et al. OipA "on" status of *Helicobacter pylori* is associated with gastric cancer in north-eastern Brazil. BMC Cancer. 2019;**19**(1)

[112] Lima VP. *H. pylori* (CagA) and Epstein-Barr virus infection in gastric carcinomas: Correlation with p53 mutation and c-Myc, Bcl-2 and Bax expression. World Journal of Gastroenterology. 2008;**14**(06):884

[113] Morgan E, Arnold M, Gini A, Lorenzoni V, Cabasag CJ, Laversanne M, et al. Global burden of colorectal cancer in 2020 and 2040: Incidence and mortality estimates from GLOBOCAN. Gut. 2023;**72**(2):338-344

[114] Jiminez JA, Uwiera TC, Douglas Inglis G, Uwiera RRE. Animal models to study acute and chronic intestinal inflammation in mammals. Gut Pathogens. 2015;**7**(1):29

[115] Poh AR, O'Donoghue RJJ, Ernst M, Putoczki TL. Mouse models for gastric cancer: Matching models to biological questions. Journal of Gastroenterology and Hepatology. 2016;**31**(7):1257-1272

[116] Hayakawa Y, Fox J, Gonda T, Worthley D, Muthupalani S, Wang T. Mouse models of gastric cancer. Cancers (Basel). 2013;**5**(1):92-130

[117] Planchez B, Surget A, Belzung C. Animal models of major depression: Drawbacks and challenges. Journal of Neural Transmission. 2019;**126**(11):1383-1408

[118] Czéh B, Fuchs E, Wiborg O, Simon M. Animal models of major depression and their clinical implications. Progress in Neuro-Psychopharmacology & Biological Psychiatry. 2016;**64**:293-310

[119] Sharma S, Rakoczy S, Brown-Borg H. Assessment of spatial memory in mice. Life Sciences. 2010;**87**(17-18):521-536

[120] Liu MY, Yin CY, Zhu LJ, Zhu XH, Xu C, Luo CX, et al. Sucrose preference test for measurement of stress-induced anhedonia in mice. Nature Protocols. 2018;**13**(7):1686-1698

[121] Carvalho C, Herrmann K, Marques TA, Knight A. Time to abolish the forced swim test in rats for depression research? Journal of Applied Animal Ethics Research. 2021:1-9

[122] Wang Q, Timberlake MA, Prall K, Dwivedi Y. The recent progress in animal models of depression. Progress in Neuro-Psychopharmacology & Biological Psychiatry. 2017;**77**:99-109

[123] Réus GZ, Fernandes GC, de Moura AB, Silva RH, Darabas AC, de Souza TG, et al. Early life experience contributes to the developmental programming of depressive-like behaviour, neuroinflammation and oxidative stress. Journal of Psychiatric Research. 2017;**95**:196-207

[124] O'Mahony SM, Hyland NP, Dinan TG, Cryan JF. Maternal separation as a model of brain–gut axis dysfunction. Psychopharmacology. 2011;**214**(1):71-88

[125] Enqi W, Jingzhu S, Lingpeng P, Yaqin L. Comparison of the gut microbiota disturbance in rat models of irritable bowel syndrome induced by maternal separation and multiple early-life adversity. Frontiers in Cellular and Infection Microbiology. 2021:10

[126] Nozu T, Miyagishi S, Kumei S, Nozu R, Takakusaki K, Okumura T. Lovastatin inhibits visceral allodynia and increased colonic permeability induced by lipopolysaccharide or repeated water avoidance stress in rats. European Journal of Pharmacology. 2018;**818**:228-234

[127] Deng Y, Zhou M, Wang J, Yao J, Yu J, Liu W, et al. Involvement of the microbiota-gut-brain axis in chronic restraint stress: Disturbances of the kynurenine metabolic pathway in both the gut and brain. Gut Microbes. 2021;**13**(1)

[128] Vodička M, Ergang P, Hrnčíř T, Mikulecká A, Kvapilová P, Vagnerová K, et al. Microbiota affects the expression of genes involved in HPA axis regulation and local metabolism of glucocorticoids in chronic psychosocial stress. Brain, Behavior, and Immunity. 2018;**73**:615-624